GROW BAG
GARDENING

To my mom, who has done more for me than I can ever repay.

Inspiring | Educating | Creating | Entertaining

Brimming with creative inspiration, how-to projects, and useful information to enrich your everyday life, Quarto Knows is a favorite destination for those pursuing their interests and passions. Visit our site and dig deeper with our books into your area of interest: Quarto Creates, Quarto Cooks, Quarto Homes, Quarto Lives, Quarto Drives, Quarto Explores, Quarto Gifts, or Quarto Kids.

First Published in 2021 by Cool Springs Press, an imprint of The Quarto Group, 100 Cummings Center, Suite 265-D, Beverly, MA 01915, USA.
T (978) 282-9590 F (978) 283-2742 QuartoKnows.com

Cool Springs Press titles are also available at discount for retail, wholesale, promotional, and bulk purchase. For details, contact the Special Sales Manager by email at specialsales@quarto.com or by mail at The Quarto Group, Attn: Special Sales Manager, 100 Cummings Center, Suite 265-D, Beverly, MA 01915, USA.

25 24 23 22 21   2 3 4 5

ISBN: 978-0-7603-6868-8

Digital edition published in 2021
eISBN: 978-0-7603-6869-5

Library of Congress Control Number:
2020949445

Design: Bad People Good Things LLC
Cover Image: Kevin Espiritu

Printed in USA

GROW BAG GARDENING

The Revolutionary Way to Grow Bountiful Vegetables, Herbs, Fruits, and Flowers in Lightweight, Eco-friendly Fabric Pots

KEVIN ESPIRITU
of Epic Gardening

COOL
SPRINGS
PRESS

CONTENTS

INTRODUCTION

My name is Kevin Espiritu, and I didn't grow up as a gardener. In fact, gardening as a hobby and eventually a vocation came to me later in life, after I'd graduated from college without a clue what I wanted to do "for a living."

As a kid, I was a voracious reader and always was obsessed with science. I collected rocks, bugs, and coins obsessively. I started to grow my own crystals at home and was constantly in and out of science and outdoors camps over summer breaks.

Somehow, gardening never entered my world until I reached adulthood. Perhaps it was giving me time to explore other, less gripping hobbies before it consumed my focus entirely . . . we may never know. Whatever the case may be, the art and science of gardening have become some of the primary lenses through which I view life.

From gardening, we can learn about physics, chemistry, and biology—the foundational elements of our scientific understanding of the world. At the same time, the world of design, art, and aesthetics

Grow bags were a staple of my original front yard urban garden due to space constraints.

@EPICGARDENING

is available for us to play with. And this is saying nothing of the practical benefits we get *out* of the garden for our own personal health and happiness.

It almost feels denigrating to call gardening a hobby, for I can't think of many other hobbies that provide so many avenues to the common threads of a human life. Needless to say, it's captivated me and it's the mission both of myself and my company, Epic Gardening, to spread that mission to as many people as possible in an engaging, practical way.

ABOUT THIS BOOK

When I wrote my first book, *Field Guide to Urban Gardening*, my goal was to take you from a complete beginner in the art and science of gardening to successfully growing your own food, no matter where you live. I wanted to convey the underlying principles of gardening, so that instead of being a "cook" following a recipe, you could become a true "chef," creating beautiful, productive gardens from scratch using nothing but your own experience and knowledge.

Grow Bag Gardening takes a slightly different approach. In this book, we'll explore the world of grow bags in depth, squeezing every bit of practicality, creativity, and yield out of them as possible. Grow bags offer incredible benefits and customizability for gardeners in any living space, but grow bags find particularly strong applications in the world of small-space gardens.

This book assumes you have some basic gardening knowledge and seeks to provide you with as many options for practical grow bag gardening as possible. That being said, many of these topics are transferable to container gardening in general, as well as the wider world of gardening. If you distill the knowledge within these pages and connect

it to your broader understanding of how to grow plants, you'll get much more out of this book than how to grow amazing, healthy, vibrant plants in grow bags.

In chapter 1, I lay out the case for gardening in grow bags. Then, in chapter 2, we'll cover the wide range of options available, including making your own bags or repurposing found materials.

In chapters 3 through 5, we'll dive deep into the best plants to grow in grow bags, including plants you'd never consider, like shrubs, bushes, and even fruit trees. More importantly, we'll discuss many different soil, watering, and support methods to grow perfect plants.

High up on an urban balcony, a small amount of green space is still being cultivated.

Drip irrigation is the best way to keep your grow bag garden well watered.

Natural trrellis and support options abound when you use bamboo designs.

Turning a concrete patio into a productive garden space is simple with large grow bags.

A makeshift grow bag made from a sack can be used to grow potatoes.

Even a small urban balcony can be filled with grow bags producing a ton of food.

A large 100-gallon (380 L) grow bag functions almost like a portable raised bed.

A variety of herbs and flowers in bags with folded lips to add stability.

Letting plants sprawl out of your grow bags gives a more natural look to the method.

Grow bags are one of my favorite methods for creative small space edible gardening.

CHAPTER 1

WHY GROW BAGS?

Grow bags might seem like yet another type of container to grow plants in, but that couldn't be further from the truth. Their portability, flexibility, and air pruning benefits effectively delete the downsides of growing in containers that many gardeners face. For this reason, they've become a mainstay of my urban garden, even as I grow in the ground and in raised beds. In this chapter, you'll learn what makes grow bags so special as well as a few considerations on safety and portability.

ROOT CIRCLING: THE CURSE OF CONTAINER GARDENING

If plants could talk and you asked them how they felt about being grown in containers, there's a good chance you'd hear them whine and grumble about it. Sure, containers are one of the most versatile ways to grow a ton of food in small spaces, but we must remember that most plants don't *want* to grow in them. They're used to growing in the ground, where they evolved and adapted to grow for millions of years.

When a plant's root system reaches the edge of a common plastic or terra-cotta pot, it hits a pocket of nutrient-rich water and continues to grow—but there's nowhere for it to go. Instead of growing out farther, it starts to circle around the pot, eventually forming a pot-shaped rootball.

This leads to decreased growth rate and transplant success. An overgrown mass of roots won't establish well when moved into the garden, as the roots won't extend into the surrounding soil to mine it for oxygen, water, and precious nutrients.

Many gardeners combat a rootbound plant by popping it out of its pot and chopping off any overgrown root tissue. It seems intimidating, as most of us are used to pruning the shoots of our plants but not the roots. However, this is an effective way to deal with a plant that's excessively rootbound.

By slicing off circling roots at the bottom of the pot as well as the outer roots on the sides of the pot, a plant can be given another shot at a productive life. While this method works, some plants are less happy when their roots are pruned, and you also run the risk of damaging the plant beyond its ability to recover.

In a perfect world, we wouldn't have to do any severe root pruning because we'd avoid the problem in the first place—but how? Enter the grow bag.

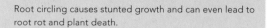
Root circling causes stunted growth and can even lead to root rot and plant death.

This response by the plant produces a branched, well-structured root system with much more surface area, giving the plant more access to the oxygen, water, and nutrients it needs to continue to grow well.

Air pruning helps prevent a plant's root system from becoming rootbound and extends the length of time a plant can be grown in a container before it needs to be repotted or transplanted into the garden.

AIR PRUNING: THE GIFT OF GROW BAGS

Due to the porosity of the material used to make grow bags, plants grown in them don't suffer the same fate as traditional container plants. When a plant's roots reach the edge of a grow bag, those roots encounter an air-rich environment that is low in water and nutrients. This causes the tips of the roots to die, signaling the plant to produce new roots elsewhere in the root system.

If a plant is being grown in a light-colored grow bag, its roots will also die due to "light pruning," as the material doesn't block out light. Prolonged exposure to light, in conjunction with the air pruning mentioned, will kill root tips as well, stimulating more balanced root growth.

In a nutshell, grow bags eliminate the number one downside to growing in containers.

THE MANY BENEFITS OF GROW BAGS

While air pruning is the principal benefit to grow bag gardening, there are a few other pluses to using grow bags versus traditional plant pots.

Versatility

Most grow bags over a certain size come with built-in handles for easy transport, which makes moving a 10- or 25-gallon (38 or 95 L) grow bag full of plants and moist soil an absolute breeze. This portability allows you to maximize production in your garden, as you can move plants around to chase the sun throughout the season.

A perfect example is potatoes grown in grow bags. Before potatoes sprout, they don't need constant

A healthy root system spreads out like a network, unconstrained by container edges.

Arranging bags from taller plants to smaller helps all plants access the light they need to grow.

Sturdy reinforced handles make moving even the largest grow bags an easy task.

Grouping bags together helps protect them from wind and cold.

access to sunlight. If your full-sun gardening space is limited, you can start potatoes in a partial to fully shaded location. While they're starting to sprout, grow another crop of radishes or quick-growing leaf lettuce in the sunny spot. By the time that crop is harvested, you should have bags with young potato foliage ready to bask in the sun's rays.

This works in the reverse too. If you need warm soil to get a plant going, start it in your sunny place and then move that bag to the shade once the plant germinates. While the soil may cool down a bit once it's no longer in direct sunlight, the plant should be able to adapt and continue to grow.

Because each bag is air pruning the plant or plants inside of it, you don't have to worry about tight quarters as you would in a more standard raised bed. This means you can crowd plants together a bit more tightly than you normally would. Be sure there's still airflow between the plants, of course, or you run the risk of plant diseases, but you can fit a lot more plants into a compact space this way.

Another fantastic way to save space is to arrange your grow bags based on plant height. Place the tallest plants behind the medium-size plants, which are behind the shortest plants. That way, all three types of plant receive consistent sunlight. This is especially useful if you're putting your plants up against a fence or wall.

When overwintering your plants, the ability to move them closer together is a major plus as well. Let's say you have a cluster of citrus trees to move into the garage for the winter. Tightly grouping your grow bags together saves space, plus it also means that they'll all have equal access to whatever grow lights you're providing.

Last, but certainly not least, is the bag itself. When it's not in use, a grow bag can fold up into a tiny package. You won't have huge stacks of terra-cotta or plastic planters piled up somewhere. All your cleaned and dried grow bags can be stored in a box and then pulled out when you're ready to use them again.

Watering

Being more porous, grow bags often require *more* watering than the average container. This sounds like a major downside, but the number one killer of most container plants is actually overwatering, which is unlikely to happen in a grow bag.

Traditional pots need to be watered less but are always at the risk of being overwatered, especially if the pot doesn't have drainage holes. With some of the creative watering techniques outlined in chapter 5, you'll learn how to perfectly manage watering your grow bags.

Temperature

As the season reaches its peak, summer temperatures can wreak havoc on dark plastic pots, heating up the soil and root zone to levels that can kill your plants. Grow bags help regulate soil temperature due to their breathability. You can also select lighter-colored fabrics that reflect heat if you're growing a particularly heat-sensitive plant.

Storage

Once the growing season ends, most pots need to be cleaned, stacked, and stored. They're heavy and can take up a lot of room, especially if you have a sizable garden. Again, the malleability of grow bag materials offer better storage, as even the largest bags can be rolled up into compact sizes for bulk storage.

DOWNSIDES TO GROW BAGS

It wouldn't be fair of me to tout the many benefits of grow bags without giving you a heads-up on potential issues, though I think most if not all of these are mitigated with proper planning and grow bag selection.

When done with your bags for the season, they're easy to clean and store in a compact way.

Longevity

On average, a grow bag will last from two to six seasons. The life span of a grow bag depends on the quality of the bag you purchase as well as the environmental conditions to which it's subjected over the seasons.

Terra-cotta pots can last *much* longer, but they are also inherently brittle. I've broken dozens of them over the years, and I'd bet you've done the same. They're inexpensive to replace, but it doesn't feel great to constantly spend money on them.

Plastic pots can last ages, provided they're made from a thicker, UV-resistant plastic. Thinner nursery pots can rip, bend, and even *melt* in hot conditions. Ask me how I know. These days, my plastic pot usage has decreased to found materials only—pots I've picked up from nurseries when shopping, 5-gallon (19 L) buckets, and the like.

Excess moisture can cause algae and mold, which is easy to clean off with a brush.

Cost

At the time of writing this book, grow bags are slightly more expensive on average than a similarly sized plastic pot. However, they're way less expensive than a glazed ceramic, metal, or wooden container. Hopefully as their popularity increases, their cost will go down. As you'll likely replace them every few years, these costs can add up. In my opinion, however, the costs are well worth it for the multitude of benefits you'll receive as a gardener.

Aesthetics

This is subjective, but many gardeners don't prefer the aesthetics of grow bags. I'm partial to them myself. The various colors, shapes, and sizes all add up to a geometric, practical look to the garden that I appreciate. If you want your pot to "do the talking," then grow bags may not be for you. I prefer my plants to do the talking and beautify my garden.

Water Consumption

As mentioned in the benefits section, grow bags lose water from all sides, meaning your plants will need

to be watered more often. This means you need to keep a closer eye on them, and you may end up using more water than you'd like. However, in chapter 5 we'll go over some strategies to mitigate this problem to the degree you can almost entirely eliminate it.

Staining

Grow bags need a little assistance when placed on surfaces that you'd like to protect from damage or staining. Because the entire bottom of the bag is permeable, a wet spot under the bag will exist more or less the entire time you're growing in it.

If you're growing on the ground in your garden, this isn't a big deal at all—in fact, you could place your bag over a persnickety weed and smother it, earning yourself a few creative gardener points. But if you're growing on a wooden deck, apartment balcony, or concrete surface, I recommend looking into some of the watering and elevation options described in chapter 5.

Mobility

Wait, how can mobility be a benefit *and* a downside? While smaller grow bags are easy to work with, bags that are 50 gallons (198 L) or larger should be considered immobile after filling them with soil, planting, and watering. The handles, if there are any, have trouble supporting the weight of all of that moist soil. While they *are* movable, I recommend waiting until the soil is on the drier side and enlisting the help of a friend. This method is the only way I'm able to move the largest grow bag that I grow in, which is a 100-gallon (380 L) bag.

Again, mobility is of less a true downside than it is something to take into consideration when deciding where to place your grow bag.

PLASTIC RESIN IDENTIFICATION CODES

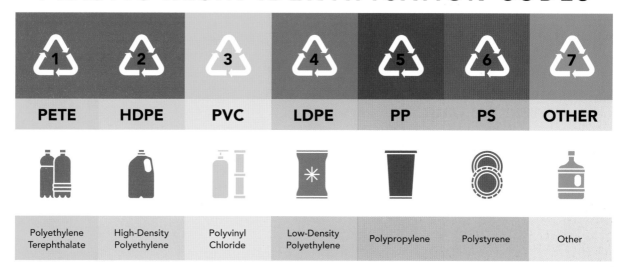

PETE	HDPE	PVC	LDPE	PP	PS	OTHER
Polyethylene Terephthalate	High-Density Polyethylene	Polyvinyl Chloride	Low-Density Polyethylene	Polypropylene	Polystyrene	Other

ARE GROW BAGS SAFE TO GROW IN?

We're all wary of plastics, particularly those of us who are eco-friendly. Many grow bags are constructed of different types of plasticized materials. Further, we use all sorts of other plastic products, from pots to old water bottles. Are *they* safe in the garden?

There are seven basic types of plastic used across all industries. Let's go over each type and what it's usually used for. Each type of plastic is marked with a number, so we'll go in order of those numbers.

1. PET plastic: PET stands for polyethylene terephthalate. This flexible material is easily the most-recycled plastic, as it makes up our water bottles, soda bottles, and other food storage containers. After long exposure to light and heat, it can begin to break down. While these can be used as a single-season mini-greenhouse, it's best not to use these for long periods.

2. HDPE plastics: High-density polyethylene makes up everything from milk jugs to drip irrigation lines. This type of plastic resists UV rays and is very heat tolerant. Most plastic pots are made out of this material because of its durability in the garden. It's a great choice.

3. PVC plastic: Polyvinyl chloride is often used in the garden because it's inexpensive. But PVC contains phthalates, chemicals that help keep the plastic flexible and durable. Unfortunately, phthalates are being studied as a possible health risk for humans. At the moment, they're generally recognized as safe in the United States, but if you have alternatives to use, go for the alternatives.

4. LDPE plastic: Low-density polyethylene is often used as plastic sheeting for greenhouse siding, or in some types of flexible food-safe containers. Like HDPE plastics, LDPE plastics are extremely heat and UV resistant, and can be an excellent choice for garden use.

5. PP plastic: Polypropylene plastic is another common greenhouse sheeting material, but it's are also used for more rigid items. While it's not as tolerant of heat as HDPE or LDPE is, it's generally recognized as safe in garden use.

The tightly woven fabric
provides rigidity while still
allowing air to permeate.

6. PS plastic: Polystyrene is extremely common. Styrofoam packing peanuts, plastic silverware, and many to-go containers are made of it. It's very common to see this repurposed for garden use. Unfortunately, this plastic isn't very sturdy, and you may end up with plastic bits in the garden. It's fine to use an old polystyrene rotisserie chicken box as a mini-greenhouse for starting seeds indoors, but try to avoid direct sun exposure that will weaken it.

7. All other plastics: Anything marked with a "7" is something to skip. The most common plastic types to find in this category are polycarbonates and polylactides, and some of these contain BPA, which is known to be hazardous to humans. Avoid these in the garden.

What Plastics Are Grow Bags Made Of?

Most grow bags are constructed of what's called "nonwoven fabrics." As a general rule, these are either recycled PET plastics or new PP plastics made of spunbond fabric. They resemble felt both texturally and visually. Spunbond polypropylene is becoming the standard for most fabric pots as it decomposes a bit slower than PET when exposed to heat and light.

Since polypropylene is usually considered safe for garden use, it's fine to use your grow bags until they start to break down. Once they lose their structural integrity and start developing weak points, it's time to replace your grow bags.

PET grow bags are very rare, although you can sometimes find grow bags with a "window" in them. That window is usually constructed of PET fabric whether or not the bag fabric is. It's best to skip these types of grow bags because the window will crack after a season or two of use, and there isn't any great reason to have a window in a grow bag in the first place.

The Final Verdict

As long as you're using a quality grade of commercial grow bag that's thick enough to hold up to the weight of your soil, it will be safe to use. These spunbond fabrics will air-prune the roots of your plants, allow ready drainage without the need to add holes, and can handle the weight of the soil with ease.

If you're DIY-ing it, stick with fabrics constructed of HDPE, LDPE, or PP for your bags. Make sure they're porous to allow excess water to escape and allow air into the soil. See the project on page 28 for instructions on how to sew your own grow bag.

You can use other cloth types to make DIY grow bags, but it depends on the material. Heavy-duty duck canvas is an option, but it will start to decay after a year or so. Canvas also tends to develop mold or mildew more readily than the plastics do. A burlap bag can look fantastic in the garden, but it will fall apart by the end of the season. Other natural materials such as jute or hemp act much like burlap does and can rapidly decay.

What about just using a plastic garbage bag? Well, you can use it, but you've got some problems. These bags are usually incredibly lightweight and won't hold up to the weight of heavy soil. You'll need to stab lots of drainage holes, and that weakens the plastic even more. Finally, these usually won't survive more than a single season. The minute you try to lift it to dump the soil out, it'll fall apart in your hands.

PROJECT: BUILD A WHEELED GROW BAG DOLLY

Once any container is filled with damp soil and plants, it starts to get heavy. If you're moving your grow bags, or any other containers for that matter, you can spare yourself from getting a backache by putting them on wheels. It'll be much easier to move them around, and you're less likely to risk damage to the plant too.

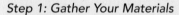

Step 1: Gather Your Materials

- (1) 1x3-inch (2.5 x 7.5 cm) cedar lumber, 6 feet (1.8 m) long
- (4) caster wheels
- 1¼-inch (3.1 cm) exterior wood screws
- Tape measure
- Pencil
- Saw
- Drill

Begin by getting your materials together. You'll need a 6-foot (1.8 m) piece of 1x3 (2.5 x 7.5 cm) lumber, preferably in cedar. This design is for a 1 square foot (30 cm²) dolly, so if you're working with a larger diameter grow bag, scale your lumber needs up accordingly. Cedar is a great choice as it holds up to the elements better than many other types of wood do.

Step 2: Cut the Boards

Cut your board into six equal-size pieces. Each one should be 1 foot (30 cm) in length. Lay two of the pieces parallel to each other on a solid work surface, roughly 12 inches (30 cm) apart. These two will be rails to support the other four boards, and you'll also be attaching the casters to these rails. You'll then place your surface boards on top, starting by making a square and then placing two boards in the center, evenly spaced. Use a pencil to mark where the boards need to be.

Step 3: Assemble the Dolly

Now that you've marked placement, it's time to assemble. Flip everything over so that it's lying on the dolly's top surface, and make sure it's all still aligned. Be sure that your boards are aligned and begin drilling pilot holes and securing the center surface boards to the two rails with wood screws. Be careful that your boards are lined up based upon your pencil marks before securing them.

Step 4: Affix the Casters

Check the positioning of your two outer surface boards and make sure the corners are aligned. Then set a caster in place on each corner of the rails. Use your pencil to mark the screw holes in the casters.

Then remove the casters and drill pilot holes at each of your pencil marks.

Step 5: Final Assembly

Place the casters back in place, being sure not to shift your boards. Using your wood screws, secure the casters to the base of your dolly. Those screws will also attach your outer surface boards. Flip it over, and you've got a plant dolly ready to use. You can sand, stain, or paint it if you'd like, or use it just as it is.

CHAPTER 2

CHOOSING THE RIGHT GROW BAGS

Like most containers, grow bags come in all different sizes. *Unlike* most containers, they also come in all different *shapes*. The flexible fabric construction means grow bag manufacturers can create square bags, deep bags, shallow bags—the sky is the limit.

Most grow bags are sold by the volume of soil they hold, but I find it's important to know the dimensions of the bag as well to help visualize where each bag will be placed in my garden.

Here's a chart from a popular grow bag manufacturer, showing the volumes and dimensions of their grow bag range:

As a rule of thumb, sticking to the 4-plus gallon (15-plus L) range is wise. Any smaller and the soil volume is so low that the bags dry out quickly and require watering far too often. Twenty-five gallons (95 L) and above is ideal for larger shrubs, trees, or root crops such as potatoes. Other plants simply don't need that much soil, which starts to get costly at that volume.

SIZE	HEIGHT (INCHES/CM)	DIAMETER (INCHES/CM)
1-gallon (3.8 L)	8.5 (21 cm)	6 (15 cm)
2-gallon (7.6 L)	9.5 (24 cm)	8 (20 cm)
4-gallon (15 L)	10 (25.5 cm)	11 (28 cm)
5-gallon (19 L)	10 (25.5 cm)	12 (30 cm)
7-gallon (26.5 L)	12 (30 cm)	13 (33 cm)
10-gallon (38 L)	14 (35 cm)	14.5 (36 cm)
25-gallon (95 L)	18.5 (46 cm)	20 (50 cm)

This style of bag is perfect for a crop like potatoes, which don't need a lot of height. More width means more potatoes.

Multi-pocket bags aren't limited to one type of crop, as you can see here with this marigold and lettuce

SHALLOW AND WIDE VS. TALL AND NARROW

Sometimes we may want to grow a lot of one crop that doesn't require much soil depth to grow well. Onions, garlic, beets, turnips, and radishes are all fantastic examples of such crops. For crops such as these, a shallow but wide bed may be the best option. The most extreme option here is a 100-gallon (380 L) bag that's only 12 inches (30 cm) deep, but about 4 feet (120 cm) in diameter, making it the perfect solution for a bed of homegrown garlic or other root veggies.

On the other hand, sometimes we want to devote a grow bag to one deep-rooted plant and condition it for future transplanting into the garden. In these cases, a deeper and taller grow bag is my recommended option. It helps maintain a healthy rootball for quite some time, allowing you to grow fruit trees in grow bags far longer than would normally be possible.

MULTI-POCKET BAGS

Similar to those classic terra-cotta herb planters that have multiple openings in the sides of the pot, these multi-pocket bags unlock unlimited planting creativity. A cut flower grow bag, herb bag, or even a multi-variety lettuce bag is easily achievable by planting each variety in its own pocket.

POP-UP RAISED BEDS

The most versatile and exciting style of grow bag is what I call the "pop-up raised bed." These come in dimensions similar to typical raised bed gardens: 2 x 2 feet (60 x 60 cm) all the way up to a 4 x 8 feet (120 x 240 cm). If you're a renter or don't want to commit to a complicated raised bed build, these types of grow bag raised beds are a fantastic option.

All you need to do is buy one, unfold it, fill it up with soil, and you're ready to plant and grow some incredible food.

This grow bag trough planter is a perfect example of a pop-up raised bed.

USING FOUND MATERIALS AS GROW BAGS

A natural inclination is to repurpose items lying around your house as a grow bag. After all, if it's bag-shaped and it can hold growing plants, it fits the definition, right? True and not true. Many found materials can be used as grow bags, but the trade-offs are usually in longevity and build quality. Here are a few I've tried, as well as my results.

BURLAP SACKS

These are inexpensive, made from an organic material, and can hold a lot of soil. They make a great single-season potato bag, for example. However, when they're wet, they tend to mold and decompose quickly, especially on the bottoms of the bags, which are in constant contact with the moist ground. This is a major problem if a bag breaks down during the growing season, which has happened to me.

POTTING MIX BAGS

These are possibly the simplest "no-work" bag of all time. Cutting a square out of the side of a bag of potting mix and then planting directly into it works surprisingly well. You don't have to mix any soil, your container is prebuilt, and you can get it up and growing in a matter of minutes.

As for downsides, potting mix bags are a bit unsightly from an aesthetic point of view, and you can also easily tear or poke the plastic just by nature of working in the garden. Drainage is also a concern, as these bags are made from plastic and have minimal (or no) aeration holes poked into them.

If you're in a massive hurry to get your garden started or feel intimidated by the idea of creating your own soil mix, then this method is a great first step to growing. But keep reading—the goal of this book is to give you every tool necessary for effective grow bag gardening.

REUSABLE GROCERY BAGS

The effectiveness of grocery bags varies with the materials used to construct them. Natural materials such as jute or canvas function similarly to a burlap sack, albeit with a bit more durability. They make a fantastic repurposed grow bag for a season or two at most.

The plastic-derived bags are effective, but there are better uses for them. And some studies have even shown excessive levels of lead that were used in the manufacture of the bags. For these reasons, I avoid using plastic-derived grocery bags for growing and opt to use them for their intended purpose.

Growing in burlap or old sacks is a perfect upcycling option for a budget-conscious grow bag gardener.

This reusable grocery bag is happy growing citronella geranium, giving the garden a nice scent.

Cutting one side of a potting mix bag open and planting directly in it is the ultimate in quick-setup grow bag gardening.

PROJECT: HOW TO SEW A GROW BAG

If you're able to sew, why not DIY your grow bags? Whether you want a square or round bag, large or small, there're plenty of methods to create them yourself.

Thread is also essential. Pick a heavy-duty polyester sewing thread. Avoid cotton threads as they will break down faster in the sun and moisture.

If you have a sewing machine or serger, you can absolutely use it. If not, a good old-fashioned hand-sewing needle is a necessity. Some straight pins can help you hold it together until it's sewn in place.

Let's go over the components you're going to need first, and then I'll give you two basic patterns you can scale to whatever size you want.

ROUND OR OVAL GROW BAG

This is the easiest type to make, and it takes very little time too. First, you'll need to decide two things: the diameter of the bag and how tall you want it to be.

MATERIALS
- Heavy-duty polypropylene landscape fabric
- Heavy-duty polyester sewing thread
- Sewing needle
- Straight pins
- Flexible measuring tape
- Scissors

The fabric you choose is the most essential part of a grow bag. I recommend a heavy-duty polypropylene landscape fabric.

For simplicity, let's say you want a bag that's 12 inches (30 cm) wide and 12 inches (30 cm) tall. Begin by measuring a 13- to 14-inch (33 to 35.5 cm) circle

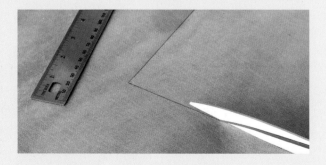

on your landscape fabric. You can use an old bucket, another pot, or even a regular ruler to help make the pattern, but you want it to be at least 1 inch (2.5 cm) wider than you need to account for your seams. Cut out that circle; this will be the base of your future pot.

Take your flexible tape measure and make a loop out of it that traces the outside of your fabric circle. Note that measurement, because that will be the length of the side fabric you'll need. Then measure a piece of fabric that's that length and 14 inches (35.5 cm) wide. Cut out this long rectangle. The extra width of the fabric will be used in making your seam and the top hem of the pot.

Use the pins to secure your long rectangle along its long side to the outside of the circle. Gather the excess fabric in the center of the circle. You can use a rubber band to keep it gathered together once you've finished pinning the edges together. Then sew all the way around the outside of the circle. You can use a whipstitch to encircle the edge of the fabric if you're hand-sewing. If you're using a sewing machine, use a tight lock-stitch about ½ inch (1 cm) from the edge of the fabric. With a serger, go with a binding stitch that's about ½ inch (1 cm) from the edge of the fabric, and don't be concerned if it trims off a little excess.

Once you get back to the start of the circle, put the two sides of the rectangle together. You should have a little overlap to make a seam up the side. Trim off the excess (if any), and then sew up the side of the bag.

Turn the bag right-side-out; if you want, you can make a nice rounded edge at the top by folding the edge into the bag and seaming it again. Then it's time to fill it up and use it.

SQUARE OR RECTANGULAR GROW BAG

What if you want a square bag? Or perhaps you want to make a long rectangle? It's essentially the same process as described, except that you have to create corners.

To do this, start at one corner and sew together your long piece and the bottom of the pot, going in neat, straight lines. Carefully bend the fabric after reaching each corner so that you can keep following the base around. Once you reach the beginning again, seam up that side.

Turn the pot right-side-out and take a look at it. If it appears square or rectangular, you're set. If it does not, turn it back inside-out. Pinch each corner together and run a straight seam up the side of the bag at each corner to form the shape. Then flip the bag right-side-out again. It should now form a neat square once it's filled with soil.

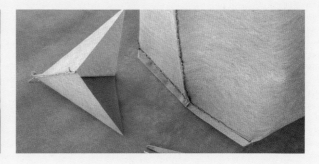

A smorgasborg of grow bags growing a grocery store's worth of produce.

CHAPTER 3

WHAT TO GROW IN GROW BAGS

While you *can* grow any crop on earth in a grow bag, your success rate will skyrocket if you select the right crops, cultivars, and bags. This is where the hard work of the plant-breeding community comes into play. Plant breeders are constantly developing cultivars suited for small-space compact gardeners to provide more options in every growing environment.

RETHINKING PLANT SPACING IN GROW BAGS

Plant spacing is a topic of much confusion among both beginner and seasoned gardeners alike. It doesn't help that seed companies list conflicting information on their packets. Most of us are growing in containers, raised beds, or, in our case, grow bags. Row-based spacing described on the backs of many seed packets doesn't make much sense in light of modern small-space gardening methods.

My gardening mentor, Mel Bartholomew of Square Foot Gardening fame, came up with an ingenious spacing methodology that informed much of my early gardening experience. In his method, he divided garden spaces into square feet, then calculated the amount of each type of crop that would fit at maturity in a single square foot. This led to a more densely planted garden, increasing yields and demystifying much of the journey for beginner gardeners.

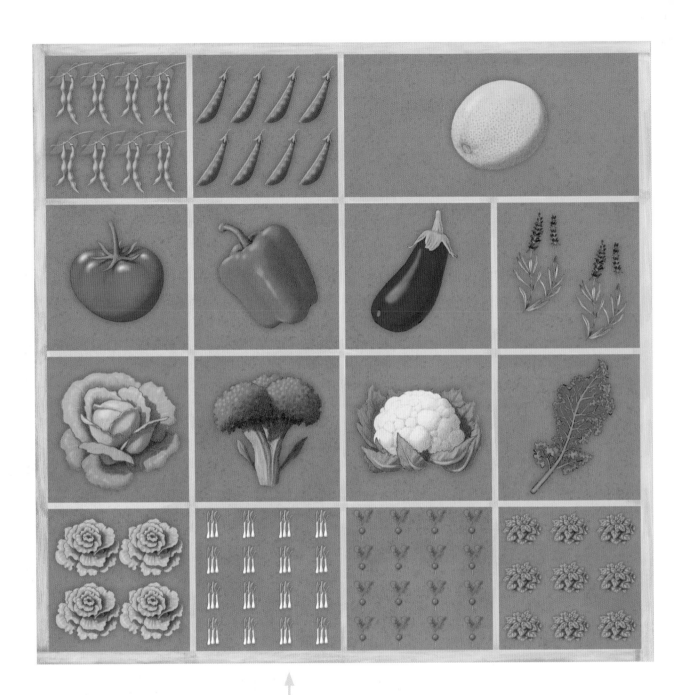

The Square Foot Gardening spacing method demystified gardening for many beginner growers, myself included.

With this methodology, someone with zero gardening experience could cross-reference his chart and know *exactly* how many crops to plant in a given space. The most common spacings were 1, 4, 9, or 16 plants per square foot (929 cm²). Although I have moved away from strict Square Foot Gardening rules, it's a method that I still hold near and dear to my heart due to my time with Mel Bartholomew.

In trying to adapt a system like this to grow bags, we're immediately presented with a problem: Grow bags offer circular growing spaces, not square or rectangular ones. This spacing is harder to deal with, as circular areas are harder to divide into neat planting grids. To solve this, use the chart to the right as a general guide to how many plants will fit in grow bags based on the volume of soil each bag holds.

Keep in mind, these are merely guidelines. They don't represent the multitude of possibilities for plant combinations, creative interplanting techniques, or even intensively planted bags. As a grow bag gardener, it's your job to take these recommendations and develop them to fit your exact needs in your growing zone and space.

Another consideration is the way you lay out your plants in the bag, which may be rectangular, triangular, or free-form in its design.

BAG PLANTING EXAMPLES

SIZE OF GROW BAG	PLANTING EXAMPLES
1-gallon (4 l)	Single small herb plants such as sage or oregano
2-gallon (7.5 L)	Larger herbs such as parsley or rosemary
3-gallon (11 L)	1–2 small herb plants, green bunching onions, leaf lettuce, garlic
5-gallon (19 L)	2 full-size onions, 4–5 garlic, 1 bell pepper, 1 bush bean
7-gallon (26.5 L)	1 determinate tomato, 1 large pepper, 6 pole beans/peas, 1 head lettuce
10-gallon (38 L)	3 seed potatoes, 5 large onions, 1 indeterminate tomato, 2 bush beans
15-gallon (57 L)	4 seed potatoes, 1 indeterminate tomato, 1 cucumber, 1 summer squash
20-gallon (76 L)	6–7 seed potatoes, 2 bell peppers, 1–2 melons (vining), 1 berry plant
25-gallon (95 L)	**2 melons, 2 cucumbers, 2 pumpkins (vining), 2-year-old fruit tree sapling**
30-gallon (114 L)	2 blackberry plants, 1 highbush blueberry (mature), 1 grapevine
45-gallon (170 L)	3-year-old fruit tree, 3 raspberries, 4 bush beans, 15–17 seed potatoes
65-gallon (246 L)	3 bell peppers, 2 grapevines, 2 tomatoes, 5 bush beans, 3 cucumbers
100-gallon (378 L)	65 garlic bulbs/leaf lettuce starts, 40 full-size onions, 4 bell peppers
150-gallon (568 L)	4 tomatoes, 4 melons, 9 bush beans, 75 green bunching onions
200-gallon (757 L)	20–25 romaine lettuces, 90 garlic bulbs/leaf lettuce starts, 4 squash

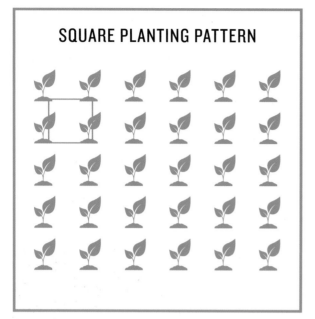

SQUARE PLANTING PATTERN

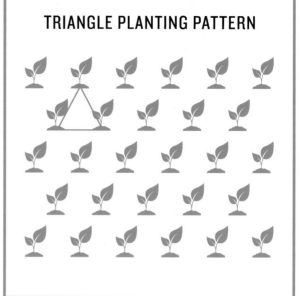

TRIANGLE PLANTING PATTERN

With rectangular spacing, divide your grow bag into a grid and plant on the intersections of the grid. This looks neat and tidy, but often won't line up perfectly with the edges of the bag.

I prefer rectangular spacing when growing in the ground or in large raised beds, as I'm less concerned about maximizing every inch of space.

Triangular spacing is slightly different and involves offsetting each row so that every plant is at a 45-degree angle from every other plant. Overall spacing is closer together in this type of layout, which is an advantage if you're trying to plant your bags as fully as possible.

Of course, you have the option to plant as free-form as you like. There's no rule stating that you must pack your bags as full as is humanly possible.

Three different apple cultivars are in one box: 'Granny Smith', 'Gala', and 'Golden' apples.

WHAT IS A CULTIVAR?

Gardening terminology can seem opaque at times, and the word "cultivar" is a prime example. A cultivar is a *cultivated variety* of a plant, meaning specific human intervention led to the development of that particular plant. For example, there are over 7,500 cultivars of *Malus pumila*, or the common apple. Among them are superstars you'll see at your local grocery store, such as 'Granny Smith', 'Golden Delicious', and 'Fuji'.

Almost all fruits, veggies, flowers, and herbs have cultivars bred for specific purposes, not just the fact that they work well in containers. For example, varieties of tomatoes exist that have a heavier fruit set, fruit earlier in the season, or have a boosted resistance to early blight. The world of plant cultivars runs deep, and I encourage you to explore it.

Maximizing Single Plant Bags

Imagine you have a bush bean planted in a 5-gallon (19 L) grow bag. You want to maximize your yield out of this bag, but you're limited to one plant, right? Not so much. While it's true that cramming more beans in would likely hamper yield, you can pepper in other plants as the beans come to fruition. Crops such as lettuce, green onions around the perimeter, or even a "groundcover" of microgreens are all fantastic options to squeeze more yield out of a smaller grow bag.

A feature plant like this shishito is surrounded by flowers to add color and variety.

VEGETABLES

Veggies, my favorite category of crops to grow, are perfectly suited for grow bags, but variety selection is critical. As I've mentioned, there is a wide variety of cultivars for most commonly grown plants, and veggies are no exception. When in doubt, look for compact varieties.

With a crop such as tomatoes or cucumbers, this means a plant that has more of a bushy, globular growing habit than a vining habit. Specifically with tomatoes, determinate varieties work quite well in grow bags, though indeterminates can work with clever trellising, which we'll discuss in chapter 5.

With root crops, the major consideration is that the total length of the root should be shorter than the height of your grow bag. For example, most radishes will work, but it will be challenging for larger varieties such as daikon to get to their full length in even the largest grow bags, as daikon radishes can grow to 24-plus inches (60-plus cm) in length.

A gorgeous array of edibles dressing up a back porch.

'Parisian' carrots grow in a compact ball shape, making them perfect for shallow grow bags.

I can't stress enough how much variety selection matters in grow bags and container gardening in general. If you choose an unsuitable variety, you'll only realize that mistake when you're too far into the growing season to correct it. Because gardening is a slow process, the choices you make early on "lock you in" for the entire season of growing. Make sure you do the prep work up front to make your garden a success.

This chart shows a few tried-and-true veggie varieties for grow bags.

VEGETABLE	VARIETIES TO TRY
Tomato	'Tom Thumb', 'Sweetheart of the Patio', 'Sunrise Sauce'
Squash	'Bush Baby', 'Starship', 'Bush Delicata'
Cucumber	'Patio Snacker', 'Arkansas Little Leaf', 'Bush Champion'
Bush Beans	'Contender', 'Purple Queen', 'Dragon's Tongue'
Peas	'Little Marvel', 'Oregon Sugar Pod', 'Peas-in-a-Pot'
Potato	'Red Norland', 'Yukon Gold', 'All Blue'
Onion	'White Lisbon', 'Walla Walla', 'Italian Torpedo'
Garlic	'California Early', 'Music', 'Nootka Rose'
Radish	Any radish variety except daikon
Pepper	'California Wonder', 'Jungle Parrot', 'Cupid'
Eggplant	'Hansel', 'Rosa Bianca', 'Fairy Tale'
Lettuce	'Chicarita', 'Little Gem', 'Monte Carlo'
Spinach	'Persius' hybrid, 'Woodpecker', 'Space' hybrid
Kale	'Dwarf Blue Curled Vates', 'Black Magic', 'Lacinato'
Broccoli	'Happy Rich', 'Montebello' hybrid, 'Monflor'

PROJECT: GROWING POTATOES IN GROW BAGS

Potatoes are my favorite crop to grow of all time, and they just so happen to thrive in grow bags. Because potatoes are grown from seed potatoes planted a few inches (centimeters) underneath the soil, they take some time to establish roots and send out shoots. During this time, you can save space in your garden by storing sprouting potatoes in a cool, dark area. Once they sprout, roll the sides of the grow bag down so the tender young shoots have better light access, and move them out into the garden.

My massive 80-pound (36 kg) harvest of red, white, and yellow potatoes from grow bags and other methods.

Step 1: *Buy certified seed potatoes, if possible, because they're produced specifically to use for growing other potatoes. Wait for them to "chit," or begin sprouting, then fill the bottom of your grow bag with 3 to 4 inches (7.5 to 10 cm) of soil. Place the seed potato in the bottom, the sprouts facing up. Use one potato per 5 gallons (19 L) of bag size. Cover with 3 to 4 inches (7.5 to 10 cm) of soil and set in a protected area of the garden.*

Step 2: *Once you see sprouts beginning to peek out of the soil, roll the sides of the grow bag down so they can access more sunlight. Every time the shoots grow 6 inches (15 cm) above the grow bag, fill with more soil until only 2 inches (5 cm) of the shoots are above ground; roll up the sides of the grow bag accordingly.*

Step 3: *Your potatoes may begin to flower, which is a sign that you have new potatoes growing underneath the soil. At this point, you can dig down to harvest some new potatoes or let them continue to increase in size. When the top foliage begins to yellow and die off, your potatoes are nearing their maximum size. Dump out the bag and get harvesting.*

Step 4: *After harvesting, let the potatoes sit in a cool place for a day or two. This allows their protective skin to harden. After hardening, you can wash off any remaining dirt to enjoy in the kitchen immediately or store for later. When storing potatoes, a cool, humid, and dark place is best to prolong shelf life.*

Grow bags are fantastic for crops like carrots, because you can customize your soil mix to match what they like.

Peppering in a few root crops around a stunning ornamental gets you a harvest with your flowers.

Root Crops in Grow Bags

It's a sad truth that root crops are left out of many gardener's containers even though they're perfectly capable of growing well in pots. The misconception that root crops need an enormous amount of space underneath the soil to spread out is the likely culprit behind their lack of selection.

In fact, there's a major benefit to growing root crops in grow bags: soil selection. When it comes to a classic root crop such as carrots, many gardeners struggle to harvest a carrot that's nice and straight, with no signs of disease, forking, or other odd issues that crop up from poor soil preparation.

In grow bags, you can customize your mix to be perfectly suited to a finicky crop such as carrots, which prefer a loose, sandier soil. At the same time, variety selection in your root crops will take you from a failed harvest to a beautiful, productive bag of delicious roots and tubers.

GROWING METHOD #1: INTERPLANTED AND BORDER ROOT CROPS

As the plants growing in your bag mature, consider planting extra root crops in the undercanopy. Plants such as tomatoes, cucumbers, and peppers will all eventually grow to the point where they reveal the soil surface in your bag once again, and their root systems are developed enough to be able to handle some extra crops planted in the bag. Roughly halfway through their growing cycle, consider inter-planting crops such as shallow-growing radishes, onions, and beets.

When growing carrots, give them a dedicated bag as their taproots don't like a lot of interference when growing.

VEGGIE	RECOMMENDED BAG SIZE	VARIETY
Radish	5-gallon (19 L)	'Watermelon', 'French Breakfast', 'Cherry Belle'
Carrot	10-gallon (38 L)	'Oxheart', 'Chantenay', 'Parisian'
Onion / Allium	10-gallon (38 L)	'Crystal White Wax', 'White Sweet Spanish'
Beet	10-gallon (38 L)	'Little Ball', 'Golden', 'Bull's Blood'
Potato	25-gallon (95 L)	'Norland', 'Yukon Gold', 'Fingerling'
Sweet Potato	25- to 50-gallon (95 to 189 L)	'Vardaman', 'Portio Rico'

GROWING METHOD #2: DEDICATED GROW BAGS

If you're obsessed with a particular root crop and want to harvest a large amount, it makes sense to dedicate a grow bag specifically to it. There are also root crops that aren't *technically* roots, but which need a lot of space underneath the soil to grow well. Both potatoes and sweet potatoes fall into this category and should be grown in their own bags in most cases.

When selecting a bag, both the depth and the diameter of the bag should be taken into consideration, especially with root crops. A wider diameter means you can cram more plants in the same bag, while a deeper bag allows a wider selection of varieties. The chart at right shows a few recommendations as to crop, recommended bag size, and variety.

FRUITS

Growing fruit might seem like overkill in a grow bag, but with the right varieties, grow bag size, and soil mix, you can grow abundant amounts of fruit. The key here is to realize that very few plants we consider "fruits" from a culinary point of view grow as annuals, so we must adjust our grow bag selection and soil mix as well as the way we care for this category of plants.

Almost all common fruit trees can be cultivated in grow bags, and some even thrive more as a result.

A cute way to repurpose avocado seeds is to pop them into a small grow bag and give them out to friends and family.

Growing Fruit Trees

Beyond air pruning, there are other benefits for fruit trees grown in grow bags.

Weather management is an option when you're dealing with any container-grown tree. If the winter gets nasty, you can move it indoors. When the heat soars, you can move it to a shadier place. There's no need to be concerned about the specific micro-climate the tree is in, as you can just shift it around. And while a potted tree is heavy, grow bags don't weigh as much as traditional planter pots do.

Containers also help regulate the size of the tree. Since it can't grow an expansive root system, it won't get too large to handle. And, because it's naturally smaller in size, you can plant more varieties of trees in a smaller space. You don't have to worry about roots becoming entangled or trees fighting one another for their food because they're each encapsulated.

But they're not without drawbacks as well. You don't get as much fruit from a container-grown tree as you would from an in-ground tree. In part, this is because of its smaller size, but it's also due to the smaller root space and lower quantities of food and water.

Remember, grow bag trees can't simply search for added nutrition on their own. They're limited to what's in the soil in their fabric pots. You'll need to be more consistent about feeding and watering to ensure that your tree thrives. Without regular feeding and watering, your tree's going to fail, so you have to be diligent to keep it alive.

Grow bags do break down over time, despite our best efforts. You may need to repot your tree into a new grow bag every few years. In addition, the weight of a tree in a grow bag can be more than those handles can manage to bear, so you'll probably want to figure out some form of plant dolly if you plan to move it regularly (see the project on page 20 to make your own).

Finally, if you're living in a climate with cold winters, you may want to figure out a way to insulate the sides of the grow bag during the winter if you're not bringing your tree inside. This will protect the roots from the effects of the colder air. Even placing other grow bags filled with nothing more than soil around the base of the tree can help provide added protection from the elements.

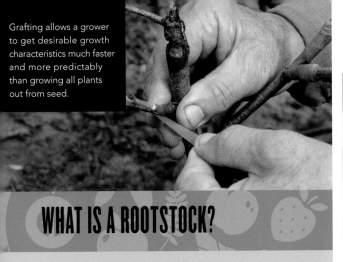

Grafting allows a grower to get desirable growth characteristics much faster and more predictably than growing all plants out from seed.

WHAT IS A ROOTSTOCK?

To preserve the beneficial traits of popular varieties of fruit, most fruit trees are grafted to what is known as "rootstock" instead of being grown from seed. If they were to be grown from seed, they wouldn't be "true to type." An avocado tree grown from the seed of the wildly popular 'Hass' avocado wouldn't be another 'Hass'; it'd be its own new variety of avocado. This might be a good thing, as the 'Hass' variety was discovered this way. But it's a multiyear gamble on what might end up being a poor variety, which most gardeners don't want to risk.

To get around this problem, fruit tree growers take cuttings of proven varieties and graft them to rootstocks, which act as the surrogate root system of the tree. This process allows a grower to clone a specific variety of fruit tree repeatedly without ever risking genetic drift.

Rootstocks themselves also impart valuable qualities to the overall tree. Some are drought-tolerant or have specific resistances to pests or diseases. Even further, some rootstocks are created to produce dwarf or semi-dwarf fruit trees, which grow significantly less tall than the "normal" variety. These dwarf fruit trees are perfect choices for grow bag gardens.

GOOD FRUITS TO PLANT IN GROW BAGS

FRUIT TYPE	VARIETIES TO TRY
Raspberry	'Strawberry Shortcake'™, 'Anne', 'Glenco Purple'
Blueberry	'Jelly Bean'™, 'Northsky', 'Sunshine Blue'
Currant	'Ben Sarek', 'Consort Black', 'Red Lake'
Blackberry	'Baby Cakes'™, 'Ouachita', 'Natchez'
Strawberry	'Seascape', 'Tristar', 'White Soul'
Grape	'Early Muscat', 'Sweet Lace', 'Hope Seedless'
Melon	'Minnesota Midget', 'Bush Sugar Baby', 'Honey Bun'
Passion Fruit	'Black Knight', 'Frederick', 'Edgehill'
Apple	Starkspur™ 'Red Rome', 'Grimes Golden', 'Granny Smith'
Orange	'Calamondin', 'Trovita', 'Dwarf Washington Navel'
Lemon	'Meyer', 'Ponderosa', 'Lisbon'
Lime	'Mexican Key', 'Bearss Seedless', 'Mexican Thornless'
Fig	'Brown Turkey', 'Black Jack', 'Celeste'
Banana	'Dwarf Cavendish', 'Dwarf Cuban Red', 'Lady Finger'
Peach	'Burbank™ July Elberta', 'Intrepid', 'Redhaven'

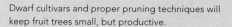
Dwarf cultivars and proper pruning techniques will keep fruit trees small, but productive.

Some trees are better than others for container-growing. While virtually any tree technically can be grown in a grow bag, there may be lots of issues to face. So it's important to pick a variety that will work well and present you with the least number of problems.

Many people adore growing citrus in grow bags. In fact, it's the number one choice. You'll need to opt toward smaller fruit sizes rather than larger ones, because huge pomelos or gigantic grapefruits won't develop on a smaller tree. Most mandarin or clementine oranges work extremely well, as do some varieties of navel orange. Virtually any lemon or lime species will thrive in a fabric pot too.

Apples, particularly dwarf varieties, can be grown in fabric pots as well. Columnar apples have in fact been bred specifically for container growing. With these, you'll need a certain number of cold hours every winter to ensure they set fruit, unlike citrus, which prefer milder or warmer weather. Be sure when choosing that you check to see what pollinators are required for your tree, as some apple species aren't self-pollinating.

Figs are champion grow bag species. Honestly, they're fantastic in containers in general, but they seem to do even better with a little air pruning of their roots. Select smaller fig varieties, and you'll have a never-ending supply of these deliciously soft, jammy fruits.

Some stone fruit trees perform better than others, but they'll all handle container-growing with relative ease. Typically, fruits such as plums or apricots that have smaller fruit sizes are more productive in a grow bag than those with larger fruits such as peaches. Cherries can be grown in a grow bag but may not fruit heavily, and they also need a certain amount of winter chill to produce fruit.

Don't forget some of the tropical trees. Dwarf banana trees are fantastic in grow bags, as are some of the more esoteric citruses such as kumquats or loquats. Unusual trees such as pineapple guava (sometimes called feijoa) also do well in containers. And of course, avocados are a must, especially if you live in a warm climate.

Avoid trees that tend to get huge or that need lots of root space. Unfortunately, this tends to rule out a lot of large-fruited varieties such as mango, pomelo, and nut trees.

How to Care for Container-Grown Fruit Trees

Begin by selecting a good candidate for container-growing. Either a bare-root tree or a potted one will work, although starting from a potted tree is easier. A tree that's one to two years of age is usually a great size to begin with.

With a bit of root pruning, this nursery cultivated plant will be perfectly healthy.

Take a look at the rootball of your tree. You'll need at least 6 to 8 inches (15 to 20 cm) of insulating soil on all sides of the tree to protect it from heat or cold. Base the size of your pot on that. For most young trees, that'll be a minimum of 20 inches (50 cm) across to start out but might be closer to 24 to 26 inches (60 to 65 cm). Your grow bag also should have some depth to it if possible, as some of the roots will grow downward.

Remove your tree from its pot, or, if it's bare root, from the bag in which it arrived. Gently open up the roots with your fingertips to make sure they're not spiraling around. Then, plant at the same depth the tree was planted before into your new pot. Don't go any deeper and be careful not to cover any grafting joints. If possible, direct the roots out toward the sides of the pot rather than straight downward. Once you've planted your tree, place a tree protector around the base of the trunk. This strip of plastic keeps mulch or other things from resting up against the trunk itself, allowing better airflow to the tree's root system. Add at least 3 to 4 inches (7.5 to 10 cm) of mulch over the soil on the outside of the tree protector, as this slows evaporation of the soil moisture and reduces weed development.

Keep mulch 1 to 2 inches (2.5 to 5 cm) away from the base of tree trunks to prevent rot or other disease issues.

Fertilize on a consistent schedule to make sure your tree has everything it needs. It may need a bit more fertilizer than it would if it were in the ground, and that's fine. You'll eventually settle into a rhythm that will work for you.

In addition, be sure to water it enough to keep the soil moist but not soggy. Some trees such as citrus need coarser, drier soil, where others prefer a moister soil type, but they all like regular watering. This may be every other day or daily, particularly during the heat of the summer.

Plants that prefer tropical climates may need to be tree-bagged over the winter to provide protection from the cold. Alternatively, you can move them indoors to shelter them from the worst of the weather. Trees such as apples or cherries that need some chill hours can stay outside, but a little extra protection for the roots can ensure they hang on even through the coldest months. Consider surrounding the exterior of the grow bag with a root-insulating layer of straw or autumn leaves, or even wrapping it in a few layers of bubble wrap.

If you anticipate needing to move your fruit tree bag, use the dolly plans from chapter 1 for easy mobility.

FRUIT TREE FOCUS: GROWING CITRUS

It may surprise you how much fruit you're able to pull off of a tree in a 10- to 25-gallon (38 to 95 L) grow bag.

Grow citrus in grow bags? It's definitely possible. Just like other forms of container trees, you can manage to get good-quality fruit out of a small space, but it takes a bit of patience to get it right. With this technique, you're able to produce a lot more trees with a much wider variety of fruit, but you'll also need to devote more attention to them.

Citrus is a wide genus, stemming from four original citrus species: papeda, citron, mandarin, and pomelo. Through hundreds of years of hybridization and plant-breeding techniques, dozens of varieties exist in all shapes, sizes, colors, and flavors. That said, most citrus varieties require similar conditions to grow well, all of which can be achieved in a grow bag.

Choose a grow bag that's at least 20 gallons (76 L) or larger for your tree so moisture levels in the soil remain stable and you don't have to repot your tree too soon.

LIGHT: With citrus, the more light it receives, the better. They prefer at least six hours of bright light, but eight hours or more is better. Locate your citrus grow bag in an area that is not blocked from the sun at any point during the day, and your tree will thank you for it.

WATERING: Citrus can survive droughtlike conditions, and in some cases, even benefit slightly from less water. That being said, never allow your grow bag to dry out completely. Water when the soil in your bag is dry to a depth of about 3 inches (7.5 cm), then give the bag a thorough soaking. As winter approaches, dial your watering back to account for a slower growth rate and lower temperatures.

FERTILIZING: When getting settled in a grow bag, citrus wants a lot of nitrogen, so an organic granular fertilizer with a high N level will work well. In chapter 5, you'll learn how to create a granular organic citrus mix that will feed your plant for many seasons. As with watering, scale fertilizing back as fall and winter approach.

PRUNING: Early on, it's a good idea to prune back fruits to limit the number the tree produces. Recommendations vary based on specific varieties, but in general a young tree should be allowed to produce five to seven fruits in the first year, and after that point the number should scale up accordingly. Also remove dead growth or any watersprouts, which are vigorously growing low shoots coming out of the tree.

TEMPERATURE: Some citrus, such as satsumas, are cold tolerant to below-freezing temperatures, but most citrus start to suffer when nighttime temperatures are consistently below 40°F (4.5°C). This is where grow bags come in handy. Using the wheeled plant dolly described in chapter 1 or some sweat equity, move your grow bags into a protected area as temperatures drop.

If you live in a colder climate, this might mean keeping them in an enclosed garage or inside your home until nighttime temperatures begin to rise again. On top of this, you may want to supplement with additional light via an indoor grow light over the cold months. Remember, citrus aren't used to winters that cold and typically would be getting ample light during that time of the season.

HERBS

The world of herbs is vast, and nearly all of them are suitable for planting in a grow bag. The beauty of many common culinary herbs is that they hail from the Mediterranean, where they benefit from drier conditions and relatively unimproved soil. What does this mean for your grow bags? They're perfectly suited to mitigate some of the drier soil conditions you might find in a bag.

That being said, there are still varieties that will perform better in a grow bag:

HERB TYPE	VARIETIES TO TRY
Mint	Common Mint, Peppermint, Chocolate Mint
Parsley	'Extra Curled Dwarf', 'Peione', 'Green Dream'
Chives	'Staro', 'Garlic Geisha'
Rosemary	'Arp', 'Gorizia', 'Prostratus'
Thyme	Common Thyme, Lemon Thyme, French Thyme
Sage	Common Sage, 'Pineapple Sage', 'White Sage'
Dill	'Fernleaf', 'Teddy', 'Hera'
Cilantro	'Santo', 'Calypso', 'Confetti'
Lemongrass	East Indian, West Indian
Basil	'Spicy Bush', 'Piccolino', 'Elidia'
Oregano	'Italian', 'Hot & Spicy', 'Greek'
Bay	*Laurus nobilis* 'Saratoga'

Mixed herb grow bags located close to the home are a fantastic way to access kitchen herbs for quick recipes.

Mint (*Mentha* spp.)

Any variety of mint will work well in a grow bag. Mint is particularly suited to grow bags due to how quickly it can take over an in-ground planting or even a raised bed. You may find that your mint quickly "bushes up" in your bag and starts to suck up a lot of water, exacerbating your watering problem. My recommendation is to use it liberally and propagate from cuttings so you can continually refresh and replace your bag.

Parsley (*Petroselinum crispum*)

Parsley is a fantastic and hardy low-growing herb that will serve your grow bags well. Whether you use it as an underplanting with a taller crop or on its own, it'll withstand drier soil and hotter temperatures than would appear at first glance. As a kid, I used to think parsley was an inedible garnish, but these days it's an absolute staple of my gardening, both in grow bags and out.

Chives (*Allium schoenoprasum*)

In my grow bags, chives serves as an easy plug-and-play interplanting herb. It works well as a groundcover of sorts, with its shallow root systems and rapid growth. The perfect cut-and-come-again herb, chives and its relatives should be in everyone's mixed herb grow bag.

Cilantro (*Coriandrum sativum*)

While cilantro remains one of my favorite herbs, it simultaneously is one of the tricker herbs to prevent from bolting. I recommend planting it in its own grow bag and placing it in a cooler area of the garden with more airflow and shade. This placement, particularly in the summertime, will slow the bolting process. However, it still wants to go to flower, and if it does, that's okay. Cilantro seeds can be saved, dried, and used as a ground spice known as "coriander." It's the ultimate double-usage herb.

Basil (*Ocimum basilicum*)

A forever favorite of mine and just about every other gardener in existence, basil absolutely thrives in grow bags. If you want to grow a basil for pollinators, I recommend 'African Blue Basil'. It develops long flower spikes and is quite hardy, so you can move your grow bag indoors, cut it back by about two-thirds, and it will come back ready to grow in the spring.

Bay Laurel (*Laurus nobilis*)

It might surprise you to learn that you can grow bay leaves, commonly used to flavor soups, in a grow bag. I find this to be one of the more exciting herbs to grow, namely because of how expensive dried bay leaves tend to be at the grocery store. With a bay laurel growing in a grow bag, you'll have a nearly endless supply of soup flavoring.

HERB FOCUS: GROWING LEMONGRASS

Anyone who likes Asian cooking is likely to be a fan of the lemongrass plant. With its lemony, slightly bitter citrus flavor and aroma, it's commonly used in some of my favorite Thai dishes.

But what few gardeners realize is that it's fairly easy to grow at home if you're in a warm climate. Even cooler-climate gardeners can grow it outdoors during the late spring and summer. It's easy to bring a small clump indoors for the cold months, allowing you to grow it year-round.

While not particularly demanding, growing lemongrass requires a few things: lots of sunlight, regular watering, and good soil. Your plants will grow to magnificent heights if you provide the right environment.

LIGHT AND TEMPERATURE

Full sun, at least six to eight hours per day, is necessary for your lemongrass plant to grow. It can tolerate partial shade if necessary, but it does best in full sun conditions.

Temperature-wise, this is a warm-climate plant through and through. Anything below 40°F (4°C) is a risk of severe plant damage and possibly death. Frost greatly damages lemongrass leaves and can kill the plant if it reaches the roots.

If you live in a cold climate, bring your lemongrass bag inside for the winter. It will readily grow as long as it has good light and consistent watering. Select a 6-inch-wide (15 cm) healthy clump to separate and transplant for overwintering.

WATER AND HUMIDITY

Don't allow the soil around your plants to completely dry out during the growing season. Lemongrass always prefers damp soil. Place about 3 to 4 inches (7.5 to 10 cm) of mulch around your plants to help keep moisture in the soil and keep the soil damp to the touch.

Plants like a little extra humidity during hot weather as well. In fact, they're extremely tolerant of high-humidity conditions. An occasional misting during dry weather will help your plants thrive.

If your soil dries out too quickly, you may want to consider placing a soaker hose under the mulch layer. This slowly seeping irrigation system helps the soil absorb moisture well without wasting water.

One of the easiest herbs to grow, lemongrass adds a ton of flavor to recipes and can be propagated endlessly.

SOIL

Loose, fertile soil is ideal for lemongrass plants. Be sure it's well draining but rich in organic material. Avoid hard-packed clay soils, as lemongrass doesn't grow well in that.

A good blend for growing lemongrass plants is two parts loamy soil to one part compost. Try to use larger-particle composts to improve soil drainage. You can add perlite if necessary, but the compost provides added nutrition.

Lemongrass isn't very sensitive to soil pH. It will grow happily in multiple ranges.

FERTILIZING

Fertilize your plants weekly. Use a half-strength solution of a nitrogen-rich, water-soluble fertilizer. Don't apply liquid fertilizer directly to the leaves of your plants. Instead, water around the bases of the plants.

This fertilization should be done throughout the summer and early fall months; June through September is recommended, as that's when the plants are actively growing. These regular fertilizing sessions will keep your plants healthy through the hot season.

If you don't have a water-soluble fertilizer that will work, aim for a slow-release granular formula. A balanced formula will work, but 8-6-6 would be ideal. This gives a little extra nitrogen to the plants to aid in stalk development.

HERB FOCUS: GROWING LEMONGRASS

PROPAGATION

Generally, lemongrass plants are propagated from seed or by division. Let's talk about these methods in more detail.

When you have a large plant, you can consider dividing it into smaller clumps. This allows a plant to have more room to continue growing, ensuring larger plants.

To do this, carefully remove the entire plant from the soil, being sure not to damage its roots. Dust off excess soil, then tease apart the clumps of stems and bulbs. You may need to use a sanitized knife to cut apart tangled roots. Replant the smaller sections and allow them to establish again. Most growers opt to divide into 6-inch (15 cm) clumps of stems and leaves.

From seed, plants take up to fourteen days to germinate. You can begin harvesting individual stalks after roughly ninety days, but it's often better to allow a plant up to 120 days, if only so it can begin to expand.

Sow seeds in moistened potting soil, just deep enough to cover the seed. Keep in a warm location, using a seedling heating mat if necessary. Once the seeds have germinated, ensure they have a grow lamp or other consistent light for at least eight hours a day.

PRUNING AND HARVESTING

With lemongrass, pruning is an ongoing process. You will want to pluck any dead material away from the bottom of your plants. Dead stalks should be removed entirely when you detect them.

As you harvest, snap the bottoms off of lemongrass stalks for replanting later.

Damaged leaves can also be pruned at any time with a clean pair of pruning shears. Be careful, as the leaf edges can be quite sharp. Wear a sturdy pair of work gloves when handling this plant.

In the early spring, before it begins active growth, a heavy trimming should occur. Trim the entire plant back to about 2 inches (5 cm) above the white part of the stalk. Overwintered outdoor plants especially need this trimming. Any leaves that have winter chill damage will be removed then.

To harvest lemongrass for culinary use, look closely at the plant. Select a healthy, thick stem, and carefully pull back the soil to expose the bulblike end. Using a sharp knife, cut the stalk off just beneath its bulb to remove the entire stalk. Be sure to get the entire bulb; if a few roots come up with it, it won't harm the plant.

My favorite tip for propagating lemongrass occurs during the harvest. If you snap the bottom of the bulb off, leaving ½ to 1 inch (1 to 2.5 cm) of bulb and root growth, you can plant those snapped ends directly back into your grow bag. It might not seem like enough plant tissue to regrow, but I've grown entire new lemongrass patches from the snapped ends of bulbs I'd harvested for use in the kitchen.

These self-seeding flowers make for an easy win if you're a new flower grower.

FLOWER FOCUS: GROWING BACHELOR BUTTONS

One of my favorite flowers, bachelor buttons are sometimes called "cornflowers." It's from this that the color "cornflower blue" derives its name.

Also called basket flowers, blue bonnets, blue cap, and many other names, this popular plant is a hardy annual. It blooms from spring through the fall in an array of colors. Leave some of the spent flowers on a plant, and it'll happily reseed into its bag and grow again next year.

Of all the flowers, bachelor buttons will be some of the easiest to grow year after year. You can fine-tune care to gain the perfect conditions, but they'll grow well even in poor conditions.

LIGHT AND TEMPERATURE

Bachelor buttons prefer full sun but can tolerate partial shade. They need lots of light to produce their array of flowers.

Temperature-wise, these plants are tough. Young seedlings can tolerate light freezes and often germinate in the late winter and early spring. Weaker plants will die back, while hardier plants continue to survive and flourish.

These plants tolerate heat, but in the most extreme heat conditions they may need extra attention to keep them healthy.

WATER AND HUMIDITY

Surprisingly drought-tolerant, bachelor buttons can tolerate everything but muddy conditions. These are not fussy plants on the whole. After all, they grow wild in many regions.

Whenever possible, avoid watering these plants from overhead. Ensure that they don't sit in areas that develop standing water either.

Cornflowers can live in humid environments, but you should provide adequate airflow to reduce the risk of fungal disease.

Moisture on the leaves can promote the formation of powdery mildew. Excessive water at the root zone can cause root rot. Do what you can to avoid these conditions.

SOIL

In its natural environment, bachelor buttons grow in loamy and well-drained soil. They're very tolerant of other soil types as well.

Also, the cornflower can be grown in neutral soils to those with quite a bit of alkalinity. A 6.6 to 7.8 pH range is favorable for these plants, although the optimal range is around 6.9 to 7.4. This provides the little bit of alkalinity they like without going too far over.

FERTILIZING

Most soil types provide ample nutrition for bachelor buttons. If you'd like to give them a boost, work in some compost or a balanced slow-release fertilizer before planting seeds.

Every month or two, spread some compost around the base of your plants. This should provide lots of nutrition and guarantee you'll have plenty of flowers throughout the year.

PROPAGATION

Sowing seed is the easiest way to propagate these plants, and it's the way that most people choose. You can sow seeds directly in the soil or start them in containers.

Once your soil's prepared, mist it to dampen the surface, and then tuck the seeds just beneath the surface of the soil. They will germinate in seven to fourteen days. Once germinated, you should thin them to one plant every 8 to 12 inches (20 to 30 cm) or so. I recommend 12 inches (30 cm) for extra growing space.

Your plants will fill in the space once they begin to really push up leaves and stems, and in time they'll flower. These flowers can produce open-pollinated seeds for next year if you'd like.

To collect seeds from bachelor buttons, wait until the flowers have faded naturally on the plants and are fully dried. Cut the flowers off its stalk, and then break it up to reveal the seeds hiding within. The seeds are oblong shaped with a tuft of brownish "hair" at the end.

Allow seeds to dry for seven to ten days in a cool, dry location, and then store them until the next planting season.

PRUNING

Bachelor buttons can become unruly if they're not supported. I'll go into that in more detail in the "Support" section.

If you choose to let them grow unsupported, too much wind can cause them to bend or lean. Trim off growth that gets in the way of paths or other plants if you wish. Aesthetics define the pruning necessity here.

Whenever, possible, try to select plants that serve multiple purposes. Here, we have beauty and pollination at play.

Deadheading your flowers is also a good choice. However, when you don't deadhead spent flowers, they will self-seed and spring right back up the next year. This also promotes more flowering.

SUPPORT OPTIONS

As I said, your plants are probably going to need some support. Since bachelor buttons can reach 3 feet (90 cm) tall, they can become a weedy tangle of bent stems if you don't offer an assist.

One of the simplest ways to support cornflowers is to place four short wooden stakes at the corners of your bag. Stretch a piece of chicken wire or nylon mesh between these stakes and use staples to anchor it in place.

As the seedlings grow, they'll stretch through the wire or mesh. This provides some low-level support against windy or rainy conditions. It also keeps small animals such as rabbits away from the base of your plants.

Marigolds beautify the grow bag garden area, while also providing pest prevention benefits.

FLOWERS

Earlier in my gardening life, I would have said to you, "If you can't eat it, why grow it?" Only looking back with a few more years of experience do I now see how foolish those words were. Apart from the sheer beauty of flowers, not only are some edible, but they all play a vital role in the overall ecosystem of a garden.

FLOWER TYPE	VARIETIES TO TRY
Sunflower	'Sunray Yellow Hybrid', 'Firecracker', 'Ms. Mars'
Rose	'Daring Spirit'®, 'Black Cherry'™, 'Royal Welcome'
Petunia	'Evening Scentsation', 'Starry Sky Burgundy', 'Spellbound White Blush'
Marigold	'Queen Sophia', 'Endurance Sunset Gold', 'Durango Outback'
Begonia	'Fragrant Falls Peach', 'On Top Sunglow', 'Vermillion Red'
Phlox	'Cherry Caramel', 'Popstars Purple with Eye', 'Scarlet'
Zinnia	'Zowie™ Yellow Flame', 'Starlight Rose', 'Pop Art Red and Yellow'
Viola	'Helen Mount', 'Amber Kiss', 'MagnifiScent® Sweetheart'
Calibrachoa	'Kabloom Denim', 'Candy Bouquet', 'Can-Can® Bumblebee Pink'
Dianthus	'Arctic Fire', 'Jolt Pink Magic', 'Chabaud Jeanne Dionis'

Dedicating grow bags to pollinator plants and placing them strategically around the garden is a great way to let nature do the work for you.

Using Grow Bags as Pollinator Attractants

I encourage you to develop a deep appreciation for the "free labor" that pollinating insects and animals provide to the garden. Instead of relying on our own efforts to pollinate all of the plants in our gardens, why not let a diverse, beautiful collection of organisms do the job for us?

That being said, I sometimes neglect to plant pollinator-friendly plants in my space-starved raised beds. This is where grow bags come in. They're perfect to plant up and pop into neglected corners and crevasses of the garden, providing food for beneficial insects and attracting more of them to your garden.

The question then becomes, "What plants are best to attract pollinators?" As with many questions in gardening, the answer depends heavily on your local climate and ecosystem. There are a few resources you should reach out to in order to get the most specific and actionable recommendations.

LOCAL AGRICULTURAL RESOURCES

There is perhaps no resource less tapped by modern gardeners than the offices of their local governmental agricultural service or, in the United States, their region's university-based agriculture extension service. Perhaps it's an awareness issue, perhaps it's a marketing one, but either way, most countries and regions have a public resource solely designed to help home gardeners with a multitude of gardening tasks:

- Identifying various plant species, pests, and diseases
- Testing your soil's pH and nutrient density
- Recommending the best pest and disease control strategies
- Recommending species and cultivars of all plant types that thrive in your area

These programs and offices are useful for far more than pollinator plant recommendations, but even if that service was all they offered, they would be worth their weight in gold. With gardening, it's vital to remember that the planting decisions we make are ones we're stuck with at least for a season, if not longer. Getting plant selection right from the start is *never* a bad idea.

NATIVE PLANT SOCIETIES

It stands to reason that the plants that are native to your region and climate are the ones that will be most beneficial to the local pollinators that frequent your area. In most cases, these species have co-evolved and have a symbiotic relationship with one another.

Within almost every city, and certainly every state (along with other countries), there are native plant societies whose sole purpose is to educate about and promote the use of more native plants in gardens.

The one I frequent, the California Native Plant Society, even has a twice-yearly native plant sale at a local park here in my hometown of San Diego. Every time a sale approaches, I get up early, grab a coffee, and roll into the plant sale with a huge cart. I inevitably overspend on dozens of native plants and then have the classic gardener's lament of figuring out where to place them in my garden. Grow bags end up being one of my favorite ways to pepper in more native plants.

If you live in North America, visit www.nanps.org/native-plant-societies to find the native plant society in your province, state, or city.

LOCAL NURSERY EXPERTS

It shocks me how often I see people at a local nursery who seem determined to figure everything out on their own, scouring the plant labels for bits of information that a nursery professional walking right by them knows *everything* about. At most nurseries, you can find at least one individual who's a lifelong gardener with both a vast and deep knowledge of what works over the course of many growing seasons.

Cherish these people as if they're made of pure gold, for they have one thing you may not: decades of experience. As I continue to hammer home, we really don't have *that* many seasons to grow in our lifetimes, especially when you compare gardening to other hobbies where results come much faster.

Tapping a nursery expert for their knowledge is one of my favorite ways to make a new friend, but also glean information that leads to dramatically more success in my gardens, pollinator-focused or otherwise.

Some plants provide pad-like flower heads that allow pollinators time to rest and get their pollination job done.

ATTRACTING POLLINATORS WITH THE RIGHT SHAPES AND COLORS

Did you know that flowers have adapted to be appealing to their favorite insect pollinators? An incredible amount of study has gone into how plants adapt to suit the species that visit them the most often, with fascinating results.

In one study, researchers opted to watch pollinator interactions with flowers in four different regions of Spain. Their results showed that, as a general rule, bees were drawn to blue or purple colors. Flies were drawn to yellow or white colors, while butterflies and moths were enticed by pink or red. Beetles favored white or cream-colored blossoms, and wasps went to browns or yellows.

UV coloring also has a role to play. Flowers that are starkly visible in ultraviolet ranges can easily attract bees or other insects, which can see light in that spectrum. For instance, some species of citrus may produce white flowers, but in UV light they practically glow. It's like providing a welcome sign for their intended visitors.

But multiple flowers of the same color didn't always draw the same pollinators. Researchers have concluded that color is not the only factor that brings an insect in for a closer look. The shape of a flower plays a large part in coaxing the pollinating insects over too.

Some flower shapes are out of reach for pollinators like bees, so hummingbirds come in to do the job, and might munch on a few pests while they're at it.

Larger flowers are more likely to draw in larger insects. For instance, the 13-inch-wide (33-cm-wide) flowers of the American lotus are the perfect landing pads for large species of winged beetles. By comparison, bees are more drawn to the smaller, but open flowers of the bird's-foot violet that allow easy access to its pollen.

Sometimes it's all about ensuring there's a landing space for their preferred visitor. A butterfly might not be interested in the aroma of a flower or its pollen but might appreciate a perch that can support its size and weight. A bee needs a little room to crawl around and do its thing, but a wasp is less likely to hang out for a while, so it needs less of a landing zone.

Tubular flowers have adapted to have long stamens upon which their pollen forms, so even these less-accommodating shapes can be pollinated by insects too. But these deeper flowers also attract hummingbirds or other species that dip their beaks into the flower. That, too, can be a method of pollination, as little bits of pollen cling to the hummingbird's long and delicate beak.

Without pollination, flowers obviously can't produce fruit or form seed. It's no surprise they've adapted their shapes and coloration to attract their favorite visitors. We as gardeners can provide a healthy habitat for bees, butterflies, and other pollinating insects by planting varieties that they enjoy visiting.

GROWING VINING PLANTS

Some of the most beloved plants in our gardens are climbers: cucumbers, beans, peas, and so forth. Unless you intentionally select a bushing variety, all of these plants require some kind of support structure to adhere to as they grow through their life cycle. Grow bags can be easily modified with a variety of DIY or purchased support structures to make almost any vining plant a possibility.

Luffa (*Luffa aegyptiaca*)

This plant is one you can grow both for its cleaning powers and to eat. The luffa, a relative of the cucumber, is harvested young as a vegetable; it has a similar flavor to summer squash. If allowed to fully mature and dry on the vine, the flesh disappears, leaving behind the dried veins of the plant. Cracking off the rind exposes those dried veins, and once the seeds are removed, the veins are an incredible scrubber. These will need a trellis to grow on, but otherwise are a fantastic addition to a grow bag.

Pole Beans (*Phaseolus vulgaris*)

All beans are good options for grow bags, but pole beans can be prolific even in a small space. Provide a tipi-type trellis for them to climb up to form a living green cover. Harvested young, most pole beans can be used as a green bean. Older ones can be dried and shelled to be stored for later use. These are great when grown between other crops in your grow bags, as they restore some nitrogen content of the soil. There's a huge variety out there, but check out 'Seychelles' as a green bean, 'Red Noodle' as a fresh-eating or shelling type, or 'Gita' as an Asian long bean variety.

Pumpkins (*Curcubita pepo*)

Most people wouldn't consider growing pumpkins in a grow bag, but they're one of the most fun crops you can grow. Unless you're growing a dwarf variety, you'll need to provide a lot of space for those vines to trail. Aim for one large pumpkin to a 10-gallon (38 L) grow bag or up to three plants in a 25-gallon (95 L) grow bag. Dwarf varieties will produce more compact vines, but they will also produce smaller fruit. Of the dwarf types, check out the 'Hooligan' or 'Munchkin' varieties for a 3- to 4-inch (7.5 to 10 cm) orange pumpkin, or 'Baby Boo' for a similarly sized white one.

Watermelons (*Citrullus lanatus*)

Nothing says summertime like a sweet, flavorful watermelon . . . and you'll want to grow them every year. Like pumpkins, watermelon vines can trail out a bit, so provide some space for non-dwarf varieties and similar spacing in the grow bag as pumpkins would get. Miniature varieties that produce those lovely little "personal watermelons" include 'Sugar Baby' and 'Golden Midget', but you can pick up one of the unusual large varieties such as 'Moon and Stars' too. Be sure these get plenty of water every day, as they'll need it.

Cucumbers (*Cucumis sativus*)

If you love pickles, you'll definitely want to grow cucumbers to make them. Pickles need to be prepared when the fruit is just barely ripe and fresh from the vine, so you'll have the best ones by growing your own. Even if you're not a pickle fan, you'll be thrilled with the amount of uses for homegrown cucumbers, from mixed drinks to salads and slaws. Check out 'Diamant' for gherkin pickles, 'Boston' for larger slicers, 'White Wonder' for an ivory-colored salad addition, or 'Marketmore' for a standard supermarket-style cucumber. For an unusual lemony tang and tiny fruit, consider 'Mexican Sour Gherkin', sometimes called mouse melons because of their appearance. They look like thumb-size watermelons.

Peas (*Pisum sativum*)

What's not to love about peas? The Asian pea varieties produce long, flat pods that are lovely in stir-fry, or you can opt for the sugar snap types that are round and juicy. If you'd prefer peas for dry storage, shelling peas can become the basis for wintertime split pea soups. They grow rapidly in the cooler months of the year, can have lovely flowers and foliage, and are almost as good at nitrogen restoration as beans are. 'Oregon Sugar Pod' is an excellent snow pea variety. I'm partial to the multicolored pods of 'Sugar Magnolia' as a sugar snap pea, and 'Alderman' is a perfect shelling pea to grow on a very tall tipi-type trellis.

GROWING PERENNIAL VEGETABLES IN GROW BAGS

Looking for something you can harvest year after year? Consider growing some perennials in grow bags. While you may need to opt for larger bags for some varieties, you can get a harvest every year from these incredible vegetables.

There's a huge variety of species available to choose from. Some are well known, such as asparagus, which may easily be one of the most common perennial

If you know you're going to move in the future, you can still grow perennials by keeping them in large grow bags.

Unbeknownst to many, when we harvest artichokes we're eating the unopened flower head of the plant.

vegetables grown today. Others, such as skirret, have fallen from favor over the centuries but are making a comeback today.

Let's go over a selection of these perennial favorites, as well as a little information about growing these in your fabric pots.

Artichoke (*Cynara scolymus*)

Perennial only in warm climates, the artichoke is a rather large plant. At full maturity, the plant can be up to 6 feet (1.8 m) across, although 4 to 5 feet (1.2 to 1.5 m) is more common. As such, you'll need a large grow bag just to support the size and weight of the plant itself.

Opt for a bag that's at least 24 to 36 inches (60 to 90 cm) across. Wider may actually be better, as it also provides more room for root spread. If you can, place this on a permanent base with casters just in case you need to move it, as it will be quite heavy.

When you're preparing for the winter after the fall harvest, cut back your plant to ground level, and cover with 4 inches (10 cm) of mulch. Straw or leaves are commonly used as mulch for this plant, but other composts are also an option.

Asparagus (*Asparagus officinalis*)

Even in cold-winter climates, gardeners can easily grow asparagus as a perennial. If you live in a warm growing zone, you will need to provide a location where it can stay relatively cool during the summer's heat.

Extremely popular for its tender shoots, asparagus crowns are a favorite for bag growing. However, like artichokes, these need a lot of space. Its roots will spread out to fill the entire space, which is desirable as they'll produce more of those shoots in the spring.

Opt for a large grow bag, something in the 36-inch (90 cm) range. This ensures plenty of space for the plant to spread, plus stability for its tall fernlike foliage.

Capers (*Capparis spinosa*)

What exactly are the little greenish round things we call capers? They're actually the flower buds of this plant, which is sometimes called the Flinders rose. They're quite bitter when harvested, but after pickling they are a fantastic addition to your cooking.

In tropical climates, these will be an ongoing perennial you'll love. This bushy plant can easily be grown in fabric bags, although if the temperature drops below 18°F (-7.8°C), you'll want to bring them indoors or protect them with a cold frame.

French Sorrel (*Rumex sculatus*)

The lemony lance-shaped leaves of French sorrel can be a fantastic addition to salads or sauces or used as a leafy cooked green. Its sharp, acidic flavor can be an excellent counterpoint to chicken or fish dishes. While sorrel is not as popular in the United States, foodies are rapidly turning to this mound-growing greenery as a new perennial option.

Those in cool climates where winter temperatures can dip as low as -10°F (-23°C) should have little trouble growing this as a perennial, although in colder climates it may die back over the winter if it's not in a cold frame. For year-round greens, move your bag into a greenhouse or cover with a cold frame in the winter. In the summer, provide a little shade, as French sorrel performs best in cooler climates.

Garlic (*Allium sativum*)

Most experienced gardeners will look at this profile and suggest that garlic is grown as an annual crop. But it's technically a perennial plant. If left in the soil, each clove on a head of garlic will resprout and grow. Once it does, you can pop it up, separate the cloves and replant them for an ever-renewing supply of garlicky goodness.

Garlic can be grown perennially in most non-tropical climates. In a grow bag it does fantastically well, so garlic is definitely a good choice. For best head development, try to provide at least 4 inches (10 cm) of space between plants.

Groundnut (*Apios americana*)

Two foods from one crop? With the groundnut, sometimes referred to as "hopniss," that's definitely an option. *Apios americana* likes to stay consistently damp, so you'll need to water this one a lot in a grow bag setting. Still, you can grow both the tasty tubers and the beanlike seedpods in climates where the lowest winter temperature doesn't dip below -30°F (-34°C) and summers aren't excessively hot.

Because it's a tuberous crop, be sure not to overplant your fabric pots. You'll want room for these to spread out and develop more tubers along their rhizomes. The seedpods can be harvested young and cooked like green beans or left to develop mature seed for planting.

Garlic is a long-season crop that, if grown correctly, will provide an abundance of bulbs for a year or more in the kitchen.

Horseradish (*Armoracia rusticana*)
Most don't think of horseradish as a vegetable, but more of a spice. Still, this sharp and head-clearing root comes from a perennial plant. In cold climates, a good coating of mulch can protect the roots through winter and allow you to keep it thriving.

In grow bags, this can be a complex plant due to the size of those roots. They get massive, and as such, you'll need a whole lot of soil for them. Opt for a 25- to 30-inch-diameter (62.5 to 75 cm) fabric bag at the smallest and try to get one that's nice and deep as well. This will give you the best chance of producing massive hot (spicy) roots.

Lovage (*Levisticum officinale*)
Its leaves are considered herbs, with a parsley-celery flavor that tastes good in broths or on salads. The root can be used as a cooked vegetable or grated into salads. And the seeds are used in a way similar to fennel. Lovage is really the everything-edible perennial you'll love growing.

In cold climates, lovage grows as a perennial, although in the colder months its leaves will die back. It does need a good mulch layer to protect its roots through winter. Provide a minimum of a 10-gallon (38 L) grow bag, and consider going larger, because this plant can get quite tall.

Radicchio (*Cichorium intybus* var. *foliosum*)
A type of chicory, radicchio grows as a perennial in cool-climate regions where winter temps stay above -10°F (-23°C). It has a much wider growing range as an annual, but this cool-weather crop overwinters well in those zones. It forms a small head, similar to a cabbage, but with a distinctly different flavor.

These grow well in grow bags but prefer cooler temperatures to warmer ones for best development. Keep the soil moist, and mulch around the plant to prevent soil moisture evaporation if the weather warms up.

Once established, rhubarb is prolific and makes a wonderful pie ingredient.

Rhubarb (*Rheum rhabarbarum*)
Hardy to -30°F (-34°C), rhubarb is a perennial that many immediately turn into delicious pies. While its leaves are inedible due to the amount of oxalic acid in them, the cooked stems are an excellent vegetable, but really become the star of the show once they're prepared for sweet dishes.

A fairly sizable vegetable, rhubarb can reach 3 to 4 feet (0.9 to 1.2 m) across, so be prepared with a wide grow bag for this plant. Harvest the side stalks as they reach maturity to prevent them from drooping over the sides of the bag.

Scarlet Runner Bean (*Phaseolus coccineus*)
Most beans are grown as annuals, but in warm climates where winters stay above freezing, the scarlet runner bean can be a perennial vining plant. It's possible to coax this bean into a mild perennial form, although it's shorter-lived than some other perennials at only a few-year life span.

Still, for beans, perennials are rare—and it does extremely well in grow bag formats. Provide a tipi-type trellis or other support for the vines to clamber up, and you'll find this to be a great addition to your perennial patch.

Shallot (*Allium cepa*)
Like its cousin garlic, shallots can be treated as perennials. In fact, they may be slightly better suited to being grown as a perennial than garlic is. For best bulb formation, it's still important to separate the cloves when possible, but you can let them overwinter in a cluster if you'd prefer.

Growing zones that don't dip below -30°F (-34°C) are best for overwintering shallots, and you can grow them in warm climates, too, if they're well shaded during the hottest part of the day. They perform wonderfully in grow bags.

Skirret (*Sium sisarum*)
A relative of carrots, skirret is relatively unknown in the United States. Elsewhere, it's treated in much the same way as similar root vegetables such as carrots or parsnips. Their roots can be fibrous if the plant is more than a year old, but young roots tend to be tender and sweet once cooked.

Skirret grows extremely well in fabric bags where winters stay above -20°F (-29°C). If you live in a hot climate, treat this as a shade plant. Divide the plant annually to ensure it has plenty of room to produce more edible roots and to harvest some for eating.

Sunchoke (*Helianthus tuberosus*)
Hardy to -20°F (-29°C), the sunchoke is a bushy plant that produces edible tubers that taste remarkably like artichoke. That might be why they're sometimes called Jerusalem artichoke. Those tubers also sprout right back up in the spring, so when you harvest, leave a few tubers behind.

They perform well in grow bags but need to be watered more regularly, as they need a lot of moisture. Overcrop your tubers with a short-season cool-weather plant in the late winter or early spring, something such as leaf lettuce or radishes, and you'll get double use out of your fabric pot.

Runner beans are perfect for a DIY bamboo trellis, which we'll cover in chapter 5.

Sweet Potato (*Ipomoea batatas*)
This vine is both beautiful and productive. Its tubers are absolutely delicious and well worth growing, especially in hot climates where it's a perennial. Not only that, but the flowers it produces are stunning, and the leaves can be eaten too.

In a grow bag, it's best to limit the number of plants to one or two depending on the bag size. This provides plenty of room for the tubers to grow under the surface, providing you with those delicious sweet potatoes. Leave room around the fabric bag to allow the vines to spread.

Walking Onion (*Allium* x *proliferum*)
The walking onion is an interesting plant. Unlike many other alliums, this performs best if it's grown densely packed. It self-propagates by bulblets off the base of the parent, so will form huge mats of onions.

Hardy to -10°F (-23°C), these are perennial, although you may have a little dieback in the colder months. Every year or two, divide walking onions so that they can continue to spread out and produce more bulblets.

STRATEGIC COMPANION PLANTING STRATEGIES

There is no more confusing topic in gardening than companion planting. Infused with bits of mysticality *and* science, it can be difficult to tell whether the planting charts circulated around the internet are based on anything at all.

It's not that specific plants "do/don't like one another" as much as it is practicality. For example, tomatoes and broccoli are both large plants with large root systems, and you shouldn't make them share space with each other. Putting them at close proximity means they're competing for soil nutrients and water, all while fighting for space. As a result, both plants will end up being weaker than they should be and more susceptible to pests and diseases.

Practical companion planting is close-quarters planting with actual reasoning behind it. Instead of having vaguely metaphysical overtones, it supports specific goals. Some plants need shade that other plants might provide; others may act as a living mulch or pest repellent near plants that need their assistance.

A prime example is the Native American tendency to plant three specific crops together. This combination, often referred to as the Three Sisters, is corn, beans, and squash. The corn's rapidly growing height provides a sturdy living trellis for the bean plants to climb. The squash vines spread out around the base of the other two plants and act as a living mulch. All three benefit from this combination.

So how can you work practical companion gardening into your grow bags? It's actually quite easy to do and can produce incredible results for both your food and your flowering species.

Basil and tomatoes is a classic companion planting, as the basil shades the soil and keeps it moist for thirsty tomato plants.

Despite what many think, corn can successfully be grown in grow bags provided you keep the soil moist.

What's Your Goal?

This is the question to answer when determining your plant combinations. Figure out what one plant *needs* and identify another plant that fills that need. Here are a few of the most common situations you'll run into.

NATURAL SUPPORT OPTIONS

Some plants grow tall and strong, such as bamboo. In fact, in chapter 5 you'll learn how to turn bamboo into a variety of different support structures, so even after harvesting it has an endless number of uses in the garden.

That being said, in companion planting we're focused on living plants providing a benefit, so I turn to crops such as sunflowers and corn as fantastic examples. Both can be planted in a bag, allowed to germinate and establish themselves, and then be paired with a plant that would benefit, such as pole beans or peas.

PROVIDING SHADE

Consider plants that need more shade than others. A good way to provide this would be to plant taller plants "in the way" of these more-sensitive plants, providing a microclimate that offers protection from hot, bright afternoon sun.

A classic example is growing lettuce or another low-growing leafy green underneath a summer crop that's about halfway through its life. For example, sowing lettuce under tomatoes is a perfect way to squeeze out some greens in the peak of summer, when they would typically suffer from the oppressive heat and light levels.

WEED SUPPRESSION AND LIVING MULCH

While weeds aren't a huge problem in grow bags, water loss can be an issue. The two main ways a grow bag loses water are evaporation and through use by the plants that are growing within it. While you can't protect from evaporation on the sides of the bag, as they provide the air pruning benefits you want, sowing an edible groundcover in a bag will protect the top from water loss.

One sneaky way to do this is by heavily sowing classic veggies and growing them as microgreens around the larger crops you've planted. I like doing this with peas, sunflowers, and arugula, but you can do this with just about any microgreen profiled in chapter 6. It's a creative way to protect the surface of the soil as well as get a quick harvest from the unplanted areas of your bag.

Another more common approach is to sow frilly crops such as loose-leaf lettuce, endive, spinach, and so forth down below to provide some shade over the surface of your bag. As a bonus, the larger plant you're planting these under is *also* providing shade for these more light- and heat-sensitive greens. It's a double whammy of practical companion planting.

TRAPPING PESTS

What about a crop that suffers heavily from an annoying pest, such as aphids? Instead of breaking out your organic pesticide of choice, why not plant a "trap crop," such as nasturtiums, in a grow bag? These beautiful and easy-growing plants act as an attractant to aphids, meaning you can lure aphids from the crops you want to protect in a natural way that builds your garden ecosystem at the same time.

Grow bags are particularly effective for this trap-cropping strategy, as you can move them around the garden strategically to place near the plants you'd like to protect. When you see aphid or other pest populations rise on the trap crop, simply cut away affected leaves and bury them in your compost, feed to your chickens, or discard. Any way to incorporate that organic matter back into the garden in a way that doesn't perpetuate the problem is my preferred choice, though.

LURING BENEFICIALS

The exact opposite case would be attracting beneficial insects to the area of your choice. This is achieved in the same manner: Choose plants that naturally lure pollinators and predatory insects to them, ideally near a plant that also needs their help.

WHAT DOES NITROGEN FIXING REALLY MEAN?

One of the most touted benefits of legumes is their ability to "fix" nitrogen from the atmosphere into a solid form, which can then be used by plants. But how does this really happen, and is it at a scale that matters to our gardens?

At a basic level, legumes such as beans and peas have nodules on their roots that act as a home for rhizobia, a family of bacteria. These bacteria feed off the sugars the root nodules provide, and, in return, they convert atmospheric nitrogen to ammonia. This ammonia is converted into nitrate, which is then used by the plant.

Here's the tricky part: Most of the nitrogen found in living legumes is contained in the stems, leaves, and seeds. As a gardener, these are the parts of the plant we're eating, especially the seeds. So while legumes do fix nitrogen from the atmosphere, we almost always remove most of that nitrogen when we harvest.

On the flip side, legumes are an often-recommended cover crop for improving poor soil. In these cases, the recommendation is to grow them and then either let winter frost kill them or chop and drop them right on top of the soil. All of the organic matter and nutrients within will break down and build the soil, including the nitrogen fixed from the atmosphere.

This works quite well, but for the purposes of grow bag gardening, it is not a practical suggestion. That being said, the remains of your bean and pea plants should make their way through some kind of composting system, where they'll be broken down and can be added back to your grow bags the following season.

The symbiotic relationship between rhizobium bacteria and legumes create these "lumps" on the root system.

GROW BAG COMPANION PLANTS

PLANT	PRACTICAL BENEFITS
Basil	Pest Repellant
Bean, Bush	Nitrogen Fixer
Bean, Pole	Nitrogen Fixer, Shade Provider
Borage	Pest Repellant, Beneficial Attractor
Calendula	Pest Repellant, Beneficial Attractor
Catnip	Pest Repellant, Beneficial Attractor
Chamomile	Beneficial Attractor
Cilantro	Pest Repellant
Clover	Beneficial Attractor, Living Mulch
Corn	Living Support
Cosmos	Beneficial Attractor
Dill	Beneficial Attractor
Echinacea	Beneficial Attractor
Garlic	Pest Repellant
Lettuce	Living Mulch
Marigold	Pest Repellant, Beneficial Attractor
Melon	Living Mulch
Mint	Beneficial Attractor
Nasturtium	Trap Crop, Living Mulch, Beneficial Attractor
Peas	Shade Provider, Nitrogen Fixer
Squash	Living Mulch
Sunflower	Living Support, Beneficial Attractor

SUCCESSION PLANTING IN GROW BAGS

Basic Succession Planting

Much like a raised bed or an in-ground garden, succession planting in grow bags is a fantastic way to squeeze more produce out of a limited space. Let's look at a basic application of succession planting by examining two scenarios:

Scenario 1: You plant lettuce in a grow bag and harvest it in fifty days, then sow another round of lettuce, and harvest in fifty days. Total days: 100.

Scenario 2: You plant lettuce in a grow bag and harvest it at fifty days, and upon harvesting, transplant a two-week-old lettuce seedling to the bag. You harvest this lettuce in thirty-five days. Total days: 85.

By starting seeds before transplanting into your grow bag, you'd save fifteen days, and this is only on two rounds of lettuce. If you could grow lettuce year-round, succession planting in this manner would yield ten full harvests of head lettuce per year compared to seven if sown from seed.

Succession planting can be confusing because you're dealing with all sorts of crops that grow at different rates, and you're trying to fit those crops into your growing season in a way that maximizes your harvest.

I find it's simpler to think in planting cycles for succession sowing based on the time it takes a particular crop to grow. Short-cycle crops can be continuously sown every one to two weeks, and long-cycle crops can be continuously sown every three to eight weeks.

SHORT-CYCLE SUCCESSION PLANTING

PLANT	DAYS TO HARVEST	PLANTING INTERVAL	GROWING NOTES
Arugula	30 days	14 days	Cool weather preferred
Bean	60 days	10–14 days	Plant in spring & summer
Beet	40–70 days	14 days	Plant in spring & fall
Broccoli	60–70 days	14 days	Plant in spring & fall
Cauliflower	50–65 days	14 days	Plant in spring & fall
Corn	70–100 days	14 days	Late spring through summer plantings
Endive	40–70 days	14 days	Plant in spring & fall
Kale	40–50 days	14 days	Plant in spring & fall
Lettuce (head)	70–85 days	14 days	Plant in spring & fall
Lettuce (leaf)	40–50 days	14 days	Cool weather preferred
Peas	55–70 days	10–14 days	Plant in spring & fall
Potato	90 days	14 days	Plant in spring & fall
Radish	25–30 days	7 days	Cool weather preferred
Spinach	50–60 days	7 days	Plant in spring & fall
Turnip	35–40 days	14 days	Cool weather preferred

Seasonal Succession Planting in a Grow Bag

In the previous examples, we looked at growing the same crop time after time. But what if you want to grow along with the seasons, while simultaneously maximizing how much you get out of your grow bags? This is where seasonal succession planting techniques come in handy.

Let's look at the case of a single bag grown throughout the seasons:

- **Late Winter:** You sow peas in your bags indoors and move them outside once the temperatures warm up.
- **Spring:** The peas are growing, you build the bamboo tipi trellis described in chapter 5 to support them, and start tomato seedlings indoors as the peas are about halfway through their life.
- **Summer:** Your peas reach the end of their life cycle, so you remove them from your grow bag, amend the soil, and transplant an approximately one-month-old tomato plant, saving yourself that month of time.
- **Late Summer:** Your tomato produces wonderfully throughout the summer and starts to taper off as the heat dies down. You've started a few kale plants about three weeks prior to removing your tomato. You transplant them into the bag and are able to harvest more fresh kale before a cold front comes in and ends the season.

LONG-CYCLE SUCCESSION PLANTING

PLANT	DAYS TO HARVEST	PLANTING INTERVAL	NOTES
Cabbage	70–80 days	21 days	Plant in spring & fall
Carrot	85–95 days	21 days	Plant in spring & fall
Cucumber	60 days	21–30 days	Late spring through summer plantings
Eggplant	65 days	30–60 days	Plant in summer
Melon	80–90 days	21–30 days	Up to 12 weeks before first frost
Okra	70 days	21 days	Plant in summer
Onion (green)	85 days	14–21 days	Spring through fall planting
Summer squash	45–60 days	30–60 days	Plant in summer
Swiss chard	60 days	30 days	Up to 10 weeks before first frost
Winter squash	90–120 days	30–60 days	Plant in summer

Peas, peppers, and Asian greens: a perfect season succession planting.

The lesson here is clear: Always be sowing. Seeds and soil are inexpensive when you compare them to the lost opportunity of not having something to plant.

By always having something ready to plant in your grow bag, there is zero downtime where a bag is empty, waiting to be planted. When growing in small spaces, making the best use of the containers at your disposal can quite literally double or triple the amount of produce you can grow in a year.

The mix you choose to fill your grow bag is the #1 factor to your growing success, so choose wisely.

CHAPTER 4

FILLING YOUR GROW BAGS

Compared to other growing methods, soil in grow bags tends to dry out faster, so our soil mix must compensate for that. We also want to increase nutrition because increased drainage can lead to nutrients leaching from the bottom and sides of the grow bag during heavy rains or watering sessions.

No matter what method you're using to grow, all soil mixes must provide the following qualities:

- **Drainage:** The mix must be porous enough to allow water to drain, preventing root rot.
- **Aeration:** The mix must have ample space between particles for pockets of air to provide roots with much-needed oxygen.
- **Nutrition:** The mix must have all the nutrients a plant needs to thrive, and in ample quantities.

THE SUSTAINABILITY OF PEAT

Peat-free mixes like the one you see here are easy to make and perform well.

One of the most hotly contested debates in the gardening world is how sustainable it is to use peat moss in soil mixes. Peat comes from bogs and wetlands, primarily in Canada and Russia. Although it's technically a renewable resource, as peat bogs do regenerate over time, many experts say that the regeneration rate is so slow that it should be considered a nonrenewable resource. Thus, we shouldn't use it as much as we do in our gardens.

In response, many gardeners turned to coconut coir—ground and processed coconut husks—as a viable alternative. But on this side of the debate, detractors say that the human and natural resources needed to process, wash, package, and ship coconut coir are as much, if not more, destructive to the environment as harvesting peat.

The verdict is unclear, but one thing is for sure: No matter which soil amendment you choose for your garden, using it responsibly without wasting it is the best course of action.

A PEAT MOSS ALTERNATIVE

If you'd like to avoid peat moss and want to build your own peat-like mixture, I highly recommend this mix from my friend Stephanie Rose and her book *Garden Alchemy*:

- 1 part compost
- 1 part OMRI listed coconut coir
- 1 part rice hulls or perlite

This mixture provides a bit of everything: organic matter, water retention, and drainage. To be honest, it could even be its own potting mix, but it works well as a substitute for peat moss. In the following recipes, feel free to mix up a big batch of this peat alternative and use in place of peat moss or coconut coir.

PERLITE VS. VERMICULITE: WHAT'S THE DIFFERENCE?

In name *and* appearance, perlite and vermiculite can confuse many gardeners. Though they seem the same, the attributes these two amendments add to the soil are different.

Perlite (at right in the photo above) is made from a specific type of volcanic glass with a high water content. It's heated, which evaporates the water into gas, thus expanding the glass into what we call perlite. Perlite is a good choice when you need to improve both drainage and aeration of your soil mixture.

Vermiculite (at left in the photo above) is made from compressed dry flakes of a silicate material that is absorptive and spongy. It's a honey-brown color, and the little flakes will expand to hold three to four times their volume in water when hydrated. Because of this, adding vermiculite to your soil improves water retention more than perlite does while also not offering as much aeration.

EPIC GROW BAG SOIL MIX RECIPE

This mix is my recommended all-purpose soil mix for grow bags. It does a great job holding enough water to keep the root zone moist and adding enough fertility to keep even the heaviest-feeding plants happy for quite some time.

When mixing this together, I recommend using a large tarp and wetting the mix as you go. Peat moss and coconut coir are difficult to rehydrate at first, but once they start to moisten you should have no problems. Mix as you go to fully and evenly incorporate all ingredients throughout.

Base Mix
- 1 part peat moss, coconut coir, or peat alternative
- 1 part perlite, pumice, or lava rock
- 1 part compost, ideally from multiple sources

Optional Additional Ingredients
For each gallon (3.8 L) of soil mix, add the following ingredients:
- ½ cup (106 g) worm castings
- ½ cup (113 g) kelp meal
- 1 cup (227 g) native soil

Soil Amendments: Three to Consider Adding to All Mixes

While there is a whole host of soil amendments to choose from when making a soil mix, there are a few tried-and-true ones I keep going back to.

Worm Castings

This is one of my absolute favorite amendments of all time due to how easy it is to generate at home. Worm castings are one of the best components you can add to spruce up a soil mix, bar none. The waste by-product of worms, they're highly active in living biology: bacteria, fungi, and even worm cocoons. The nutrients contained within worm castings, while not overpowering from an NPK standpoint, are water-soluble.

Worm castings also:

- Act as a pH buffer, keeping your soil's pH within a tighter range and allowing nutrient uptake to continue.
- Increase a soil mix's ability to retain water, a classic struggle in many containers.
- Do not burn your plants via overfertilization, as it is practically impossible to overapply them.

When you couple these benefits with the fact that worm castings are one of the easiest amendments to generate for *free* at home, you have a truly epic combination. I use worm castings liberally in all aspects of my gardening, and grow bags are certainly not excluded.

Kelp Meal

Living in coastal San Diego, California, and growing up going to the beach to fish and surf has left me with an affinity for ocean-derived fertilizers, with kelp- and seaweed-based amendments at the top of my list.

The body of knowledge around the benefits of kelp both in and out of the garden is still growing, but here's what we *do* know about its power as a soil amendment:

- It contains seventy-plus trace minerals, vitamins, plant growth hormones, and amino acids.
- It's high in potassium, and as such is fantastic for crops that demand a lot of potassium during their growth.
- It's an abundant, natural resources that can be self-harvested or purchased for little money off the shelf.

In my own gardens, I've experimented with drying kelp and creating my own kelp meal, or even composting it directly and adding to garden beds. But for the purposes of adding to a grow bag mix, I recommend purchasing an organic kelp meal product.

Native Soil

A few years ago, I would have recommended bacterial or fungal inoculants, especially when growing in containers. These products can help add life back to a mix that might be devoid of many of the organisms that make up a quality soil food web.

Anecdotally, I've experienced fantastic results from adding a sprinkling of these products to my soil mixes, specifically mycorrhizal inoculants. For this reason, I *do* recommend using them in a container mix, as it's a human-created mix that can be devoid of biological life.

However, these inoculants can be expensive, and they often only contain a few species of bacteria or fungi, when a well-functioning soil has hundreds or thousands all competing and cooperating with one another.

For this reason, I like to add a scoop of native soil directly from my yard or garden to most of my mixes, with the logic that if that soil is growing healthy plants directly in-ground, it stands to reason that the organisms within it are contributing in some way to the plant's health. Because a grow bag is isolated from native soil, adding a scoop into the mix introduces these organisms to their new environment in the bag.

Your mileage may vary here, and I wouldn't recommend adding soil you know to be infested with a soilborne pest or disease.

8 DIFFERENT SOIL MIX RECIPES

While the Epic Grow Bag Mix works well for most gardening applications, some plants require special soil conditions to grow well. When creating a custom soil mix, all you're doing is shifting the percentages of different soil variables, such as:

CONDITION	EXAMPLE
pH	Blueberries like a more acidic (4.2 to 5.5) pH than most commonly grown fruits and veggies.
Water Retention	Delicate plants such as ferns prefer soil that retains more water than cacti and succulents.
Texture	When starting seeds, a lighter, more fine-grained mix allows tender roots to take hold and develop.
Nutrients	Hungry plants such as tomatoes want a soil that's rich in organic matter.

These modified soil mixes should help you when growing plants like berries, fruit trees, cacti and succulents, houseplants, seedlings, and more. If you start thinking about making soil as adjusting for a specific plant's needs, your success rate will go up.

Heavy Feeder Mix

Many of the veggies we grow in the garden, specifically ones in the cabbage family, are what we call "heavy feeders." All that means is that they require more nutrients than the average veggie. These plants include but are not limited to: asparagus, beets, broccoli, Brussels sprouts, cabbage, corn, eggplants, kale, kohlrabi, okra, parsley, peppers, pumpkins, squash, and tomatoes.

Couple their increased nutrient requirements with the fact that they're planted in grow bags, and it's a good idea to shift the soil mixture to include more organic matter:

- 1 part peat moss, coconut coir, or peat alternative
- 1 part perlite, pumice, or lava rock
- 2 parts compost, ideally from multiple sources
- ¼ part worm castings
- ¼ part kelp meal
- ¼ part native soil

The increased compost ratio will feed these hungry plants over the course of their lives, but additional fertilizing may be necessary to supplement, which we'll cover in chapter 5.

Citrus Mix

Citrus require well-draining soil, so a normal soil mix that retains water well isn't going to work. If you already have a large batch of Epic Grow Bag Mix, you can take the easy path by "lightening" that mix up with ingredients that have larger particles, such as:

- Pine chips
- Coconut husk chips
- Coarse sand
- Perlite
- Vermiculite

Amend a classic potting soil mix with about 30 to 50 percent of these ingredients to lighten it up, selecting a few ingredients of different particle sizes to decrease uniformity and compaction.

If you want to make a citrus mix from scratch, here is my recommended recipe:

- 1 part bark fines
- 1 part perlite
- 1 part Epic Grow Bag Mix or potting soil

Seed-Starting Mix

There are a few important things to know about seeds and the germination process. Once understood, making your own seed-starting mix will make perfect sense.

First, seeds don't need any nutrition from the soil in the first week or two of their life. The cotyledons, or seed leaves, are responsible for storing and providing the nutrients that a seedling needs immediately after germination.

Second, if you're going to start seeds indoors, it makes sense to give them a mixture that makes life easy for them. By that I mean that the mixture should provide all of the signals a seed needs to start the germination process. These include ample moisture and a lack of large particles that could block a young seedling's push out of the soil and into the light.

Knowing this, here's a fantastic seed-starting mix:

- 1 part peat moss, coconut coir, or peat alternative
- 1 part perlite
- ½ part vermiculite or worm castings (if using worm castings, use slightly less than ½ part)
- A couple handfuls of fine sand
- A handful of agricultural lime (if using peat moss, to neutralize the acidity)

Mix these ingredients together, and then moisten. Spread into your seed-starting tray or grow bag, and then sow your seeds. If you're growing a plant that requires darkness to germinate, consider adding a thin layer of vermiculite on top to as a "micro-mulch" of sorts. This helps lock in moisture as well, which can be particularly handy for finicky seeds such as carrots.

Acid-Lover Mix

Some plants absolutely love an acidic soil. Blueberry is a prime example, as it does best in a pH range of 4.5 to 5.5. The easiest way to achieve an acidic pH is to buy a bagged mix specifically designed for acid-loving plants. You'll often find this at the garden center sold as a camelia, azalea, or gardenia mix, as all three of these plants love acidic soil. Pick a few bags up and fill your grow bags, and you're set.

If you'd like to try your hand at making your own acidic soil, here's a recipe:

- 1 part Epic Grow Bag Mix
- 1¼ cups (284 g) elemental sulfur per plant
- 1 cup (255 g) garden gypsum per plant

In my own grow bags, I typically opt for a bagged acidic soil mix due to the difficulty of getting soil to the right pH and keeping it there.

Tree and Shrub Mix

Trees, shrubs, and other perennials are going to be in your grow bags for quite a while, so it makes sense that they'd need a different soil mix to keep them happy. The main difference here is that we've added

well-rotted woodchips to the mix. Be sure they're rotted well, though, as woodchips mixed into soil tend to tie up nitrogen as they break down. A little sand in the mix as well helps with drainage.

- 1 part peat moss, coconut coir, or peat alternative
- 1 part well-rotted bark or woodchips
- 1 part compost
- 1 part sand
- 1 part perlite

Succulent and Cactus Mix

Succulents and cacti can be grown effectively in grow bags and even benefit from the extra drainage typical of grow bag gardening. That being said, they still need a custom mix with excellent drainage and minimal amounts of nutrition.

- 3 parts peat moss, coconut coir, or peat alternative
- 1 part perlite
- 1 part vermiculite or worm castings
- 2 parts coarse sand
- A handful of agricultural lime (if using peat moss) to neutralize acidity

A mix like this will hold just enough moisture to keep the plant alive, but all extra moisture will drain right through.

Tropical Houseplant Mix

Tropical houseplants tolerate a wide variety of soil conditions, and as such they'll grow well in the standard Epic Grow Bag Mix. However, if you're a houseplant addict like me, you may want to customize slightly for optimal performance. This mixture forgoes compost in lieu of more water retention and drainage elements, which are key for tropical plants. You'll add some organic granular fertilizer to help feed your houseplants, but you should also consider a liquid fertilizer regimen once every month during the growing season.

- 2 parts peat moss, coconut coir, or peat alternative
- 1½ parts perlite
- ½ part sand
- Handful of organic granular fertilizer

Large Grow Bag Mix

Large grow bags can be best thought of as more mobile raised beds. The sheer volume of soil and size of the bag means that water loss simply isn't as big an issue as it is with a smaller bag. If you've got lots of finely shredded woodchips or wood shavings, you've got part of what you'll need to make a good large grow mix following this recipe.

- 1 part finely shredded woodchips or wood shavings
- 1 part peat moss, coconut coir, or peat alternative
- 1 part compost or composted manure, with large chunks broken up
- ½ part perlite
- ½ part worm castings

The wood can hold extra water, as can the worm castings and compost, while the perlite allows excess moisture to drain away. You can add a little sand to this blend if it seems to need a bit of extra drainage.

There are dozens of other variations you can make to provide the perfect texture blend for specific plants. Container trees, for instance, need a lot of good drainage and perform well with some organic fertilizer blended through. Orchids like large, chunky bark that allows tons of airflow to their roots. Root vegetables do better in a finer-grained soil than in one with lots of large particles. Just mix together the components you need to generate your own soil, and you'll never run out.

BAGGED POTTING MIXES: WHAT *ARE* THEY?

It's essential to get yourself a quality potting blend before you plant. But what actually is a potting blend?

Bagged soil is a bit perplexing when you look at it at the store. Some bags will state it's for raised beds, others will be earmarked for amending poor soil. Others will say they're for seed-starting or for acid-loving plants.

Even the term "soil" may be a misnomer. Many of these contain nothing that technically could be called "soil" at all. So how do you determine which you need?

Why Do You Need Different Mixes?

Different growing conditions and different plants call for different needs. A succulent, for instance, is going to be fine in your fabric bag, and it's not going to need much water. But a tomato plant needs consistent, even moisture to protect it. And a tree is probably going to be craving water when it's leafing up and producing new fruit.

Needless to say, different manufacturers create blends that cater to specific needs. So what are some of the conditions that you might want to create?

- **Moisture retention:** Keeping the soil moist is necessary for many types of plants.
- **Good drainage:** To prevent root rot, a soil that can release excess water is key.
- **Loose texture:** Some plants like the soil to be loose and easily penetrated by roots.
- **Good aeration:** Plant roots need air to breathe too.
- **High acidity:** Acid-loving plants such as blueberries prefer an acidic soil blend.
- **Nutrient density:** Some materials break down in the blend and provide constant food.

- **"Living" soil:** Beneficial microorganisms can defend plants against disease along with many other benefits.
- **Sterile conditions:** A bagged mix is unlikely to be infected with soilborne diseases.

There are others that are even more heavily customized to specific plant needs, but this provides a good idea of why one would want a variety of options.

MOISTURE CONTROL INGREDIENTS

Certain things help regulate the moisture in your potting mix. They usually serve one of two purposes: They hold on to moisture or they let it drain away.

In the "drain away" category, there are materials such as perlite, rice hulls, and sand. Technically, sand is a soil type, so in mixes that include sand, there is some soil. When sand is in a mix, it doesn't hold moisture, so the water will flow right through.

Perlite is an expanded volcanic glass filled with nooks and crannies that hold small amounts of liquid, but they allow the majority to flow through the soil. Rice hulls are literally the hulls of rice grains that have

been stripped off during processing, and these can act as a biodegradable version of perlite.

Inputs that hold on to moisture include things such as composted manure or plant-based compost, worm castings, vermiculite, coconut coir, or forest products such as shredded bark. All of these absorb water and hold on to it until the plants need it, making them valuable additions.

TEXTURE AND AERATION INGREDIENTS

A loose, lightweight potting soil is often held up as the ideal, but it's not perfect in all situations. Some plants actually prefer a soil that is a bit denser. For most edible plants, the soil's tilth, or "dig-ability" for lack of a better word, is important, as it ensures good root penetration.

Larger materials provide more aeration. Orchids and other related plants often demand much more airflow around their roots, and the chunky material of orchid bark is optimized to their needs. But even in your cucumber bag, you'll need some larger material to promote good airflow too.

Forest products are a popular additive. Made from shredded wood or bark, these can provide some bulk to the mix to allow air to penetrate. Other aeration materials include perlite, rice hulls, or decomposing straw or hay.

But what if you need something more than a particularly coarse soil? There are lightweight options that provide a good texture to your mix. Peat moss often makes up the body of many soilless potting blends, with coconut coir a close second. Worm castings are also an excellent addition for providing good tilth, as the fine particle size keeps things loose.

A staple in my transplanting routine, mycorrhizal fungi supercharge your soil by helping plant roots access nutrients more effectively.

NUTRITION AND PH-RELATED INGREDIENTS

Some components provide nutritional benefits for your plants. The most common of these are composted manures (cow, horse, chicken) or composted plant materials. Occasionally, added nutrients such as alfalfa meal or bat guano can be incorporated to boost soil nutrition as well. Be careful when fertilizing with these types of potting blends, as they're often optimized straight out of the bag to have every-thing a plant needs initially.

Not all plants prefer a neutral soil, and so by integrating additional materials, a potting blend can raise or lower the pH. For example, sphagnum peat moss tends to be slightly acidic, and thus is used as a common addition to acidic soil blends. Elemental sulfur can also lower the pH to a more acidic range. Other components, such as calcium, increase the pH toward the alkaline side.

LIVING INGREDIENTS

Finally, we come to ingredients that are alive and active. Materials such as worm castings or composted manures can be full of densely packed microbiology. The soil microbes residing in these mix ingredients can assist your plants to draw up nutrition more readily or protect them from soilborne plant diseases.

Other sources of living microbiology are mycorrhizal fungi or beneficial bacteria that have been added to your blend. Like the microbes mentioned previously, these can provide similar assistance to your plants. Certain species of mycorrhizae will form a symbiotic relationship with the roots of your plant, sharing resources with the plant while both thrive from the pairing.

Blends that have been used year after year may have become colonized with the spores or bacteria that cause plant disease. Purchasing a new batch of bagged mix can ensure it's sterilized and free from these risky microorganisms. Sometimes they'll also be inoculated in advance with the more beneficial varieties so that plants will have the advantage of microbial boosters.

TOPSOIL: WHAT IS IT AND IS IT USEFUL?

Often from construction sites, topsoil found at big box stores is not to be used as a sole ingredient in a grow bag mix.

On occasion, a few mixes incorporate topsoil. This is literally nothing more than average dirt, finely screened to remove large particles. As an actual soil rather than a soilless option, dirt can be a worthwhile addition to many mixes as it provides a mineral base in which plants can grow.

At the same time, topsoil on its own isn't particularly beneficial. It lacks most nutrient qualities that you'll find in many of the other mix inputs. The perk of using it is that topsoil does not break down the same way that other inputs do, so it's less likely to sink down over extended use. But as it doesn't break down, you'll need to keep mixing in organic additions to provide the richness your plants require.

Commercial Bagged Mixes

Depending on the brand and the intended use, commercial potting mixes are all over the spectrum in terms of what ingredients they contain. Let's compare two different all-purpose potting mixes and talk about what's in them.

One popular company produces an organic potting mix for all plant types and containers that uses a sphagnum peat moss base of about 35 to 45 percent of the blend. Into this, they mix a wide variety of things in a proprietary recipe. Some of the common components are aged forest products and/or forest humus, perlite, agricultural lime, worm castings, alfalfa meal, kelp meal, feather meal, and yucca extract. They also add a blend of ectomycorrhizal fungi and endomycorrhizal fungi to their potting mix.

From this description, we can identify a few things about this mix. Alfalfa, kelp, and feather meals are all nutrient-related additives, as is the yucca extract, so it's already fertilized and ready to go. The fungal population that's been added will help plants absorb nutrition and stay healthy. The forest products, humus, and worm castings provide moisture retention, and the perlite keeps it light and well aerated. Since they've included agricultural lime, it's likely pH-neutral and ready for planting.

To contrast this blend, let's compare it to another brand's container mix. The second company's inputs include processed and recycled forest products, arbor fines, perlite, peat humus, sphagnum peat moss, dehydrated poultry manure, feather meal, composted poultry manure, bat guano, kelp meal, and worm castings.

In this variety, it's likely that sphagnum peat moss and the forest products make up most of the blend. But the dehydrated and composted poultry manure provides a good kick of nitrogen that will work well for most plants, while the feather meal, kelp meal, and bat guano provide the rest of a good balanced fertilizer. The arbor fines, forest products, and worm castings provide good moisture retention, and the perlite allows excess water to drain off well. This company doesn't add fungal inputs specifically, but the peat humus likely already has some in there.

This mix doesn't include the lime, but it likely is also pH neutral given that they use a mix of ingredients that's geared toward a wide range of plants. They just come at it from a slightly different angle than the first company.

The lesson here is to look closely at the back of a commercial mix's bag before you buy to make sure everything in there is something you want to add to your garden. Some brands add chemical fertilizers instead of organic ones to a few of their potting mixes, while some others are purely organic. If you know what goes into the growing medium you're using, you can be sure it's exactly what you want it to be.

Replenishing Your Soil

After a full growing season, container soil mixes are often in rough shape. They've been growing at least one crop continuously from spring through fall and as a result can be depleted of nutrients, compacted, and full of thick, fibrous roots.

In a raised bed, I typically let roots be in the soil and expect the life within the soil to slowly break them down, but in grow bags it's a different equation. Sometimes that soil needs a little extra love to revitalize and refresh it for the coming season.

Before you refresh your soil, there are a few instances in which a full replacement is recommended. If your plants suffered from a soilborne pest or disease, replacing soil is a better option.

Spent soil is full of roots, may be compacted, and in need of a nutrient recharge.

Root knot nematodes, for example, are pernicious pests that are nearly impossible to rid from affected soil without taking drastic action. It's better to throw the soil away, or plant varieties that are resistant to their effects.

Verticillium wilt and fusarium wilt are two common fungal diseases borne from the soil that affect popular plants such as tomatoes. Again, if you detect their presence in your soil, throw it away rather than reuse it for the next batch of plants.

Assuming your soil is pest- and disease-free, it just needs a little love. Here is my three-step process for refreshing your grow bag soil:

Step 1: Dump and Screen

First, empty your grow bag soil onto a clean surface, such as a large tarp. If you're refreshing multiple grow bags of soil, combine them to mitigate any nutrient deficiencies specific to one bag.

Gently work the soil through a ¼- to ½-inch (6 to 12 mm) screen. Keep an eye out for grubs or other soilborne pests.

To keep it simple, top off with more of your chosen grow bag mix, or add custom amendments based on what you want to grow next.

Remove any large chunks of debris and root tissue. Optionally, you can sieve the soil through a screen built from ½-inch (1 cm) chicken wire stapled to a wooden frame. This helps find any grubs or other pests who bury in the soil and munch on plant roots.

Step 2: Amend
The easiest way to amend an old soil mix is to add more potting mix, or whichever soil mix you custom-made for your bags. This increases water retention, breaks up any compaction, and adds more organic matter for the next batch of plants.

A good ratio is about three parts old mix to one part new mix. That minimizes use of new soil mix while still giving your containers the boost they need.

If you want to provide a more targeted amendment, consider supplementing with specific organic granular fertilizers that are heavier in the nutrients that the last crops used in high amounts. For example, tomatoes

are a heavy-feeding crop, so soil in grow bags with tomatoes as their last crop may need a heavier dose of compost or fertilizer to get them back in shape for the next round of planting.

Step 3: Add Biology and Condition
This last step is completely optional, but it can help kick-start grow bag soil mixes. Organic granular fertilizers and compost have a lot of nutrients within them that plants need, but often require other soil microorganisms to break them down into their constituent parts so a plant's roots can use them.

I add a healthy dose of worm castings produced from my worm bin to get this job done. Worm castings are a light, but balanced, NPK fertilizer, but their real benefit is the rich environment of microorganisms present to start colonizing your grow bag and breaking down the organic matter. If you're lucky, you may get a few earthworms in your bag as well.

The refreshed soil mix back in the bag, ready for another successful season of growing.

Conditioning Soil for the Next Season

If you're hanging up your grow bags for the season, this technique may help you start the next season with rich soil full of organic matter that's ready for a spring planting. Instead of emptying and folding your bags up for the winter, try leaving them full of a prepared soil mix and letting that slowly break down over the course of the winter.

As long as you keep your bags in an area that doesn't fully freeze and keep the soil relatively moist, your can follow the process for replenishing you grow bag soil mixes at the end of the season and then leave that soil in the bags over the winter. If you like, applying 1 to 2 inches (2.5 to 5 cm) of compost across the top of the bag as an additional topdress can help add even more organic matter to your bag come springtime.

Mix a few scoops of fresh worm castings (dried if you don't currently practice vermicomposting) into your bag and water in well with rainwater, or water that's been sitting out for at least twenty-four hours to off-gas chlorine, which would kill much of the soil life you're trying to cultivate.

Let this newly refreshed grow bag mix sit, keeping it watered well, for about two weeks. This waiting period gives the microbial life within the bag time to start breaking down organic matter. Feel free to plant at any time after this point.

This is a condensed version of a classic approach to building soil over the fall and winter in an in-ground garden or even a raised bed. By covering the soil with organic matter—fall leaves, compost, or even a winter-killed cover crop—you protect the surface from swings in temperature and moisture loss, and allow organic compounds to continue to break down. Come springtime, your soil is ready to go and less disturbed than if you'd made a new mix from scratch.

If you can't store your bags in a protected area, a simple frost cover like the ones in chapter 5 can help boost soil temperatures in the bag enough to keep the soil from freezing. You can also group bags so they're not exposed to the elements on all sides.

Drip irrigation is my
go-to method for easy,
automated watering.

GROW BAG GARDEN MAINTENANCE

Compared to most container gardens, grow bags require a bit more tender loving care. From proper watering techniques to strategies to prolong the life of your bags, these guidelines should help you manage your grow bag garden effectively without spending too much time or effort.

WATERING GROW BAGS

As we know by now, the price we pay for the air pruning benefits of grow bags is that they tend to dry out faster than the average container. Correct watering, aside from being the most important repeating task you do in any garden, is doubly important when growing in porous containers.

The question, "Why does everything I plant in grow bags immediately die?" is one of the most common I receive, and the answer is almost always, "Because you're watering incorrectly or not setting your bags up for success." Let's discuss some strategies for watering grow bags. These strategies will vary in terms of their complexity and effectiveness, ranging from simply watering more all the way to some cool projects you can build to keep your bags watered perfectly 24/7.

Although hand-watering with a can works, in this chapter you'll learn much more effective watering methods.

Watering Frequency

The easiest way to counteract water loss in grow bags is to water less but more often. This differs from the advice given in my previous book, *Field Guide to Urban Gardening*, where I recommend watering *deeply* but *less* often. How can this be?

Every growing situation provides different benefits and downsides and to treat our plants the same would be a mistake. In a grow bag, watering deeply but less often will lead to oversaturation for brief periods of time, nutrient leaching, and extended periods of dry soil—not a recipe for success.

Make watering your grow bags part of your morning routine. Check the soil with your finger or a moisture meter and top off each bag with a bit of water if necessary. This peaceful addition to your day also affords you the opportunity to notice any pest, disease, or growth problems with your plants.

WATER, WAIT, WATER

In any container garden, drenching your dry containers with water is a recipe to *think* you watered your plants, but in fact you likely watered the ground. The reason for this is twofold:

1. Container soil mixes often contain peat or coconut fiber, which, when bone-dry, take time to rehydrate.

2. Pouring a large volume of water on dry soil without letting the soil rehydrate causes water to run down the sides of the container or through large cracks in the soil. Once the water reaches the bottom, it drains out and onto the ground below.

With grow bags this problem compounds upon itself if you're not careful. Water will run out the bottom of the bag *and* the sides. Combat with the water, wait, water technique:

1. Make a quick pass through your grow bag garden, watering directly at the center of the grow bag with a small amount of water.
2. Move on to another bag, repeating the process until all bags are watered.
3. Come back to the first bag and give it a deeper drink now that the soil is rehydrated, and it can accept more water.

Using this method, you can be sure that the water you're giving your grow bags is actually making it to your plants.

Breaking up compacted soil allows more water to be held in the mix, while also letting roots breathe better.

Consider Aeration

As soil settles, plants grow, and the elements exert their forces on your grow bags, it's natural for soil to become compact and/or chunky. To counteract this, take a sharp pointed object such as a chopstick and insert it into the soil. Wiggle it around a bit to loosen the soil and repeat until you've broken up the large chunks and the soil is loose and friable again.

Taking care not to do massive damage to roots, this technique relieves compaction and allows more air and water to penetrate the soil. In raised beds or in-ground, this usually isn't necessary as the life within the soil does a good job of keeping soil loose, but in containers and grow bags it can be an immense help.

Adjusting Your Soil Mix

If water loss is a serious problem in a grow bag, you may want to adjust your soil mixture to incorporate more water-retentive ingredients.

Recalling the Epic Grow Bag Mix from chapter 4, we have:

- 1 part peat moss or coconut coir
- 1 part perlite, pumice, or lava rock
- 1 part compost, ideally from multiple sources

To modify this for more water retention, I recommend the following ratios:

- 2 parts coconut coir
- 1 part perlite, pumice, or lava rock
- 2 parts compost, ideally from multiple sources

This mixture shifts drainage down and fertility and water retention up. Removing peat moss as an option in favor of coconut coir is due to the poor rehydration ability of peat moss when it's completely dry. Coconut coir has an easier time accepting water when bone-dry, which is the exact problem we're solving with this mixture.

MULCHING STRATEGIES

While mulching is a popular technique for in-ground gardens, raised beds, and ornamental landscaping, it seems to be an underutilized one in the world of container gardens and grow bags. This is a massive mistake, as there's no reason that the benefits of mulching wouldn't translate to grow bag gardening.

But what mulch to use? It goes without saying to use mulch made of organic matter, but each common type has its upsides and downsides.

Microbark: Light, inexpensive, and uniform, microbark is a fantastic mulch for grow bags. Be sure to lay 1 to 2 inches (2.5 to 5 cm) on the soil surface *without* mixing into the soil, as mixing bark and soil tend to steal nitrogen from the soil, which is the last thing we want to happen.

Straw: Light in color, straw will reflect heat from the surface of your grow bag as well as hold in soil moisture. Unless you buy a preprocessed, washed, and cleaned straw, you'll need to cut it to size as the long stems may not fit in a standard grow bag. When sourcing straw, verify where it came from and what was applied to it during its growth. You don't want straw that's been sprayed with harmful herbicides that can actually hinder your plants' growth.

Leaves: Abundant and free, crushed dead leaves are one of my favorite options. Scoop up some leaves from around the yard, crush them, and apply a 1- to 2-inch (2.5 to 5 cm) layer. The only downside to using leaves is their light weight, which can cause them to blow off completely in a gust, so leave a gap at the top of your grow bag as a buffer.

a tendency to mat. If you lay too thick a layer, the clippings may start rotting and create a nasty environment at the top of your grow bag.

When mulching a grow bag, leave a small circle of soil around the stem(s) of your plants that's free of mulch. Don't butt it right up against the stem, as the moisture buildup against a stem can be a vector for disease.

Those are my favorite four mulches, but there are many others to use, each with their own unique pros and cons, as shown below.

Grass Clippings: Another free resource, grass clippings decompose quickly and add a good amount of nitrogen to the soil. Go light on these, as they're uniform and quite thin; thus, they have

MATERIAL	PROS	CONS
Hay/Straw	Lightweight, readily available, inexpensive, easy to spread	May have weed seeds, blows off in wind, can become matted
Grass Clippings	Free, nitrogen-dense, decomposes well	Has weed seeds, may try to root and spread, can become matted
Woodchips (Bags)	Easy to get, uniform size	May be dyed or treated with herbicide
Woodchips (Arborist)	Cheap or free, decomposes well, builds soil quality over time	Usually only available in bulk as a tree trimming waste product
Shredded Bark	Extremely fine, looks great	Causes splinters, decomposes slowly
Sawdust	Finely powdered, breaks down fast, may be free	May allow weed germination, can include dust from treated wood
Rice Hulls	Eco-friendly, easy to spread	Blows off, has weeds, attracts birds
Compost	Inexpensive or free, builds soil	May allow weed germination
Shredded Leaves	Free, great soil builder	Only available in fall, becomes matted
Landscape Fabric	Prevents weeds	Doesn't decompose, needs another mulch on top to retain moisture

Milk crates, apple boxes, etc. are cheap ways to dress up a grow bag.

PROTECTING YOUR GROW BAGS

One clever way to improve water retention in your grow bags is to partially encase them in an outer shell. Using a wooden or plastic milk crate protects them from being blasted by sun and wind and thus decreases water loss. The gaps in these containers still allow air pruning of roots, preserving one of the principal benefits of grow bag gardening.

You can also purposefully build a windrow or place your grow bags in an area of your garden that's more protected. Placing grow bags at the ends of your raised bed rows is a great way to allow air circulation, but dramatically reduce sun and wind exposure on the sides of the bag.

PROJECT: SELF-WATERING GROW BAGS

If watering a multitude of grow bags seems daunting, or you want to—gasp—leave your garden for a few days, automation is the approach to take. This can be done in a few different ways, starting with the easiest: self-watering grow bags.

METHOD ONE: PLASTIC DRAINAGE TRAYS

The easiest way to self-water a grow bag is to take advantage of their porosity and give them a miniature reservoir at the bottom. You can achieve this using a terra-cotta or plastic drainage tray like the one you'd use for an indoor houseplant. They come in various diameters, colors, and materials, and are quite inexpensive.

I recommend purchasing high-sided trays that you can fill with at least 2 inches (5 cm) of water. Provided the soil in your grow bag is already well saturated, 2 inches (5 cm) of water can give your plants water for at least a few days.

METHOD TWO: GROW BAG BATHTUB

If you have a grow bag army forming in your garden, filling individual drainage trays with water can be as daunting as simply watering the grow bags . . . so why not batch them all together?

This project can be achieved in a variety of ways, including something as easy as buying a kiddie pool, but I find it's more fun to create a customized reservoir for my bags.

I recommend placing all of the grow bags you want to irrigate in a rectangular layout, then measuring around them to get your cuts for this project. In this example, we'll build a 2 x 4-foot (61 x 122 cm) reservoir.

Step 2: Attach the Liner

Lay the pond liner evenly over the frame. Press it down into the corners and bottom of the frame, ensuring you depress it completely. If it helps, set weights at the corners to ensure full coverage.

Staple the liner at the top of the 2x4, then using a razor blade, trim the excess off the perimeter.

Step 3: Place Grow Bags and Fill

Place your reservoir on a level surface, then add your grow bags to the reservoir. Fill with at least 2 inches (5 cm) of water and allow the bags to wick water upward. If need be, fill with another 2 inches (5 cm).

The only thing you need to watch out for is mosquitoes using any standing water as a breeding ground. If you notice this, using a mosquito dunk made from Bacillius thuringiensis is a great way to combat their growth.

MATERIALS

- (2) 2x4 (5 x 10 cm) lumber, cut to 2 feet (61 cm)
- (2) 2x4 (5 x 10 cm) lumber, cut to 4 feet (122 cm)
- Pond liner, cut to 3 x 5 feet (91 x 152 cm)
- Screws or nails
- Wood staples
- Razor blade

Step 1: Construct the Frame

Lay your lumber out, putting the short cuts on the inside of the longer cuts. Screw or nail the frame together.

PROJECT: DRIP IRRIGATION FOR GROW BAGS

Another way to automate your watering is to set up a drip irrigation system. You can even add an automated timer for truly hands-off watering for your grow bag garden.

Before we begin this project, know that there are a lot of different irrigation fittings to choose from, and it can become overwhelming if it's your first time installing a system. Following are a few common ways you can irrigate a grow bag.

Emitters: These are your classic drip irrigation fittings that release a predetermined amount of water on a per-hour basis. Common flow rates are ¼ gallon (0.9 L), ½ gallon (1.9 L), and 1 gallon (3.8 L). Typically attached to a drip line spike for positioning, these provide a single point flow of water directly to a grow bag.

Adjustable Drippers on Stakes: My personal favorite drip equipment for grow bags, these combine a stake to affix into the soil of your grow bag and a built-in adjustable emitter that sprays in a 360-degree or 180-degree pattern. Their adjustability allows you to customize the flow rate, direction, and even force of the emitter to your specific plant or grow bag size.

I'm a particular fan of the 360-degree adjustable emitters on stakes. The ones I use most often can release anywhere from 0 to 15 gallons (0 to 57 L) per hour of water. This is handy because you can set up an entire grow bag garden on one automated timer, yet still customize the amount of water each bag gets based on the plants you're growing or the size of the bag by tweaking the output of that individual emitter.

Soaker Hoses: Porous throughout its entire length, a soaker hose is a great way to get a consistent supply of water to a plant. My preferred application to use them in grow bags is as a ring a few inches (about 7.5 cm) in radius from the center of the grow bag.

The most effective way I've found to install drip irrigation for grow bags is by using a standard drip emitter on a spike to hold it up off of the surface of the soil. Setting your bags up on drip is best when all of your bags are near each other, so you use less materials and don't have drip line running along your entire garden. Due to the multitude of ways you can set up drip, I'll outline the best way to pull from a main line into a grow bag and leave you to connect the dots to the rest of your drip irrigation systems on your own.

- Drip line splitters, elbows, and end caps
- Drip hole punch
- Drip emitters, to deliver water to each grow bag

If you're using a drip irrigation timer, connect that to your hose bib and configure the watering interval.

Step 1: Basic Setup and Materials

At a bare minimum, you need the following components, listed in order of their assembly:

- Water source, typically a hose bib
- Drip irrigation timer, optional
- Irrigation filter, optional
- Standard garden hose, to run water close to your garden
- Backflow preventer, to prevent debris flowing into components
- Hose splitter, optional
- Drip pressure regulator, to prevent bursting lines
- Hose-to-drip adapter, to convert hose line into drip line
- ½-inch (1 cm) main line drip tubing, your main source of water for the system
- ¼-inch (6 mm) drip tubing, to pull from the main line tubing

Next, connect a drip irrigation filter to your hose if you decide to use one. If you have particularly dirty water or want to remove compounds like chlorine from your water before you irrigate, a filter is highly recommended.

Next, connect a backflow preventer, which ensures that no debris or water flows backwards from the system into the hose bib.

Next, connect your garden hose to the backflow preventer, running the hose to the area of your yard where you're setting up your grow bag irrigation. You also have the option of connecting your drip system directly to a water source if you prefer not to use a hose.

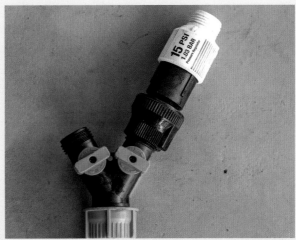

Connect your hose-to-drip adapter so you can attach your ½-inch (1 cm) main line drip tubing to the system and run that tubing out to your grow bag garden.

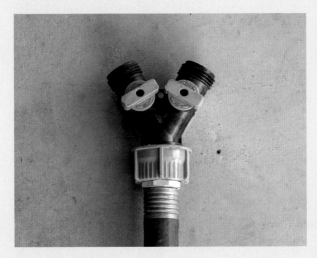

If you still want the option of hand-watering your grow bags or other areas of the garden, use a hose splitter at the water source to split the water flow and connect another garden hose to the other side of the splitter.

Connect your pressure regulator to your hose line. This does exactly what its name suggests: reduces water pressure so you don't blow off any drip attachments due to high pressure. Pressure regulators run anywhere from 10 to 60 PSI, with that number corresponding to the maximum water pressure that's allowed through the component.

Step 2: Connecting the Main Line

Lay your main line tubing out to cover every grow bag. In my example garden, I've run two T splits to feed a line of tubing down each grow bag row and used a 90-degree elbow to bump the tubing up over the garden edging I've chosen. The specific way you lay out your main line tubing will depend on the layout of your unique garden.

Cap off the main line with end caps, and then you can begin setting up irrigation for individual bags.

Step 3: Setting Up Emitters

Using a drip hole punch, punch holes in the ½-inch (1 cm) line tubing where you need to pull water into a grow bag. Using ¼-inch (6 mm) drip line tubing, hook into the main line with barbed hooks and run the ¼-inch (6 mm) tubing into the grow bag. For grow bags with a large enough diameter to accommodate multiple emitters, daisy-chain them together to create a network.

Affix drip spikes with emitters to the end of your ¼-inch (6 mm) tubing in each grow bag, centering them in the bag and placing them 1 to 2 inches (2.5 to 5 cm) above the surface of the soil. If you're using adjustable sprayers, adjust the nozzle to the right amount of water output.

Repeat the process for all grow bags you plan to set up on drip irrigation.

HOW OFTEN TO WATER ON DRIP IRRIGATION

Drip irrigation may sound daunting and overwhelming, but I promise if you play with some of the components and test it, you'll wrap your head around it quickly. At its essence, irrigation is all about getting water where it needs to go, at the right time, and in the right amounts. Once you figure that out, you'll be able to water your bags at the flip of a switch—or even automatically if you've set up an automatic timer.

The next question to follow is, "How often should I run my drip irrigation?" Like many questions in gardening, the answer is a frustrating, "It depends." Here are some factors to consider when deciding on an irrigation schedule:

After reading this chart, your mind may be spinning even further. How do you make sense of so many competing variables at once? This is where thinking like a gardener and opening up your senses to how your plants respond comes into play.

Start off with a basic schedule, such as five to ten minutes once a day, to see how your plants respond. Adjust your emitter placement and output based on specific plants and how they react to your watering schedule. If you notice garden-wide wilting, check the soil with your finger and see whether it's too wet or too dry.

Eventually you'll home in on a schedule that's perfect for the unique environment your garden is in, as well as the plants you've chosen to grow.

FACTOR	EXAMPLE
Plant Type	Cacti and succulents need less water than cucumbers and tomatoes.
Bag Size	The smaller the bag, the less water it needs to be fully saturated.
Temperature	The hotter the day, the faster water evaporates from the surface and transpiration happens within plants.
Soil Mix	Looser soils dry out faster than water-retaining mixes.
Emitter Type	The water output rate of your emitter heavily affects how long you should run your drip irrigation.
Mulch	Both the type and the depth of mulch impact water retention in soil.
Plant Maturity	Young, small seedlings need less water than a large bushy adult plant.

USING A HOSE TIMER TO COMPLETELY AUTOMATE WATERING

The Holy Grail of hands-off watering involves automatically triggering your water tap to stop and start the flow of water into your irrigation system. This is best achieved with a hose timer, but there is a multitude of different models and technologies out there. It can be a confusing topic, but once mastered it can save a ton of money, time, and, most importantly as a gardener . . . water.

MECHANICAL VS. DIGITAL TIMERS

Water hose timers come in mechanical or digital versions.

Mechanical hose timers don't use batteries. With these devices, you walk up to the hose, turn a dial to set it for however long you desire the water to be on, and walk away. It'll start the water immediately and cut it off when that set time is done.

Digital timers have multiple settings for watering options. If it's the peak of the summer and your veggies need water both at sunrise and at sunset, for instance, that's doable. These do typically require batteries, which means you'll need to replace them regularly. Easy to program, these water timers can be reset for different times of year and different watering needs.

WATERING DURATION

Depending on your garden needs, you can vary watering duration. A mechanical device is very simple: you just turn the dial and it begins the watering cycle. But if you have a multiple-outlet digital model, it's very easy to set one hose for thirty minutes at dawn and another hose for fifty minutes midday if that's what you want.

Most mechanical models operate for 120 minutes of consistent watering time as a maximum. Digital models have an even longer duration, which is great if you're deep-watering grapevines or small trees once or twice a week rather than daily.

RAIN DELAY

Clouds have formed and rain is coming. That's not a problem. All you have to do is to tell your timer to implement a rain delay for the anticipated amount of time. Now, with a mechanical model, you just don't turn the dial, but digital models vary in how this is done. Usually it's as simple as pressing a delay button or entering the number of days you want the water to remain off.

STURDY, WATERPROOF CONSTRUCTION

While it may seem silly, some poorly made hose timers aren't waterproof. While this won't hurt a mechanical model, it can destroy the whole unit on a digital hose timer. Be sure that your control panel and battery access port are fully waterproofed. It also helps if your unit is made of heavy plastic or metal, as these will hold up to the elements better.

OTHER FEATURES

If you want your garden to use wireless technology to determine when optimum watering schedules should be, there are now devices that can do that for you. With a wireless soil moisture tester, a wireless hose timer, and a base station to coordinate it all, you can let your garden go fully automatic.

For those of us who still want to maintain control over our watering frequency, there's also an app for that. Let's say that you realize while you're out of the house that it's the perfect time for your sweet potato vines to get a drink. You can use your phone app to turn on the water from afar and tell it when to turn off. Just a few taps, and it's done.

Adjusting Watering for Hot Climates

Those of us living in hotter climates often have the benefit of being able to grow longer than cold-zone gardeners, but we also have to deal with wild swings in temperature, specifically during the summer months.

To accommodate this, here are a few adjustments to consider when the heat seems determined to suck all the water out of your bags:

Water More: Adjust your watering schedule to be deeper and more frequent, favoring toward the morning over all other times of day. You want to completely hydrate your grow bags early, allowing your plants to suck up any water they need, which leaves them more able to withstand high heat as the day goes on.

Group Bags Together: The mobility of grow bags comes in handy here. By grouping your bags closely together, you'll reduce the total surface area available for evaporation to occur, and also keep a higher local humidity level in the area around the bags. Additionally, grouping allows the foliage from all bags to provide a shade cover of sorts over the entire collection of bags.

Grouping your bags together helps preserve moisture, as fewer surfaces are exposed to air.

Move to Partial Shade: If a serious heat wave is on the way, sometimes the best option is to accept defeat and move your bags into an area where they get shade cover during the hottest parts of the day. They'll still suffer from heat stress due to the temperature, but the evaporation rate will go down and they won't be subjected to blinding amounts of light and heat at the same time.

Mulch Thicker: In hotter climates, adding an extra 1 to 2 inches (2.5 to 5 cm) of mulch to the top of your bags will help prevent water loss via evaporation.

Controlling Grow Bag Drainage

If you're using the Grow Bag Bathtub strategy (see page 102), you'll keep your bags well watered while simultaneously protecting a surface from any staining as a result of constant moisture. But what if you're growing in limited space and only have a delicate surface such as wood to grow on?

The solution is simple: Elevate your bags off the surface. For larger-scale grow bag gardens, my favorite method is grabbing an old pallet from a local big box store or off one of the classified advertising websites' free section. For smaller gardens, a few cinderblocks and an inexpensive piece of 2x6 or 2x8 (5 x 15 cm or 5 x 20 cm) lumber will do.

If drainage is still a problem, it may be that you're overwatering your bags. A good method is the water, wait, water technique where you first hydrate the mulch and top layer of soil, wait for the water to hydrate the soil, and then come back with a deeper watering to moisten the entire bag. Stop watering when you see water start to drip from the bottom of the bag. Too much drainage causes issues with sensitive surfaces, but also leaches out water-soluble nutrients that you'd rather keep in the bag.

FERTILIZING YOUR GROW BAGS

Grow bags need more fertilizing than the average raised bed or in-ground garden, for a few reasons:

- Less soil volume means the soil is depleted of nutrients quicker.
- Water running out of the bottom of the grow bag can leach nutrients out of the soil.
- Plants are grown intensively and use more nutrients.

A Quick Primer on Fertilizer Essentials

Most fertilizers are sold with three numbers printed on the label, like this: 3-1-2. These refer to the percentage of nitrogen, phosphorous, and potassium present in that fertilizer. A 5-5-5 fertilizer is 5 percent nitrogen, 5 percent phosphorous, and 5 percent potassium.

This is admittedly a limited way to think about plant fertilization, but these are three macronutrients that all plants need to thrive, so the labeling convention has stuck around. Most fertilizers, especially the organic ones recommended in this book, also contain various micronutrients and trace elements plants need to thrive.

There are two ways to approach fertilizing a grow bag, both of which have their benefits and downsides. Both the plants you're growing and your own personal preferences will help guide you on the fertilization method to use for your grow bag garden.

SLOW-RELEASE GRANULAR FERTILIZER

A slow-release fertilizer is one that needs time to break down into compounds that a plant's root system can use. It might sound counterintuitive to add fertilizer to your grow bag that doesn't have an immediate effect, but there are a couple of reasons to do this.

First, if your soil mix is prepared correctly, there's ample nutrition in it to support your plants through the beginning of their lives, even if they're a heavy feeder such as tomato. Adding more fertilizer right away might be overkill.

Second, it's easier. Applying a topdressing of an organic granular fertilizer only needs to be done once or twice throughout the growing season. No fertilization schedules to manage, nothing to remember. Set it and forget it.

The downside of a slow-release fertilizer is that you can't be as precise. Because you must wait for the fertilizer to break down in order to be bioavailable to plants, there's an element of unpredictability in when that actually occurs, and thus, when plants are able to use the fertilizer.

WATER-SOLUBLE FERTILIZERS

The second option for fertilizing your grow bags is to use a fertilizer that's immediately soluble in water. Most of these options come in liquid format, which is around 98 percent water by weight. Powdered fertilizers that are water-soluble are also available, which save a considerable amount of money because you're not paying to have water shipped to you.

Water-soluble fertilizers are fast-acting, as the nutrients contained within are immediately bioavailable to plants. This makes them an excellent way to give your grow bag gardens a quick hit of nutrition as your plants mature. For example, as your tomatoes start to flower and fruit, they often benefit from a bit of extra fertilizer. Watering your tomato grow bag with a liquid fertilizer is the only reliable way of getting that nutrition to them at the exact time that they need to use it for explosive growth.

FERTILIZING DOS AND DON'TS

Under-fertilizing: Many new growers tend to under-fertilize as they think the soil mix will do all of the heavy lifting. In an in-ground garden this is usually true, provided you're building soil over time. But in containers, you need to make sure to keep on your fertilizing schedule.

WHILE DIFFERENT CROPS PRESENT DIFFERENTLY, HERE ARE SOME SIGNS OF UNDER-FERTILIZATION:

- Chlorosis
- Stunted growth
- Purple or red leaves or veins
- Dying plant tissue

Over-fertilizing: The opposite problem, tossing too much fertilizer in your bags, is a great way tp encourage the wrong types of growth, or counterintuitively stunt growth. More is not better in the world of fertilizing.

SIGNS OF OVERFERTILIZING INCLUDE:

- Leaves dropping off
- Tall plants with spindly stems
- Crispy tips of leaves
- Overabundance of foliage

Fertilizing at the Wrong Time: Both the timing in the season and the environmental conditions when you fertilize matter. As to the time of season, there are a few classic times to fertilize:

- When you build your soil mix, especially if using a granular fertilizer
- After your plants have settled in but well before flowering or fruiting
- When your plants start to flower and set fruit

As for environmental conditions, fertilizing above 85°F (35°C) is generally not recommended. If heavy rain is projected in the next few days, you may want to hold back on a liquid fertilizer as it will likely leach out after application.

PROJECT: DIY ORGANIC GRANULAR FERTILIZERS

Making your own slow-release organic fertilizers is a great way to customize the nutrients you're feeding your plants based on their unique needs. For example, a tomato grow bag will need different fertilization than a bag of leaf lettuce. While an all-purpose fertilizer will *work*, if you want to customize your fertilization further, these recipes will help.

ALL-PURPOSE MIX

This is a fantastic all-purpose fertilizer that will work well for most plants. Use ½ to 1 cup (120 to 235 ml) per plant, sprinkled right around the base.

- 3 parts blood or fish meal
- 3 parts bonemeal
- 1 part kelp meal
- 1 part chicken manure

Clockwise from the top: Bonemeal, kelp meal, chicken manure, and blood meal make up a powerful all-purpose fertilizer.

Mix the ingredients together well.

Sprinkle as needed around the base of the plants in your grow bag.

After sprinkling, water in liberally to incorporate and to kick-start the breaking-down process.

VEGAN ALL-PURPOSE MIX

This is an all-purpose mix with zero animal by-products that achieves roughly the same nutrient ratio as a standard all-purpose mix. Apply accordingly.

- 4 parts cottonseed meal
- 1 part kelp meal
- 1 part soft rock phosphate
- 1 part dolomitic lime

HEAVY FEEDERS MIX

For plants that require a lot of nutrition through-out their life span—tomatoes, corn, cucumbers, eggplants, peppers, and so forth—try this mix:

- 4 parts alfalfa meal
- 6 parts blood meal
- 4 parts greensand
- 1.5 parts kelp meal

SHRUB FERTILIZER MIX

For larger shrubs, you'll need to provide essential calcium and magnesium for optimal growth. Use ½ to 2 cups (120 to 475 ml) per plant depending on the size of the shrub, sprinkled around its base.

- 1 part kelp meal
- 1 part bone meal
- 1 part dolomitic lime

PROJECT: DIY LIQUID ORGANIC FERTILIZERS

Making your own liquid fertilizers from ingredients found in your garden or locally is a fantastic way to feed your plants with rapidly available nutrition. By taking the extra time to make DIY fertilizers in a liquid format, you also save money while making better use of the scraps coming out of your garden and yard.

VEGGIE SCRAPS FERTILIZER

We don't eat 100 percent of the plants that we grow in our gardens. Veggie scraps can be used in many ways—throwing into compost, feeding to worms, and so forth—but to make use of them quickly, this recipe works quite well.

- Veggie scraps
- Water
- Blender
- Bucket

Save enough scraps to fill about a ½-gallon (1.9 L) bucket. Throw them into a blender and purée them with water until the consistency is as smooth as possible. Pour into a 5-gallon (19 L) bucket until you're done processing your scraps and let the bucket sit overnight.

When you're ready to fertilize, add 1 quart (0.9 L) of your vegetable purée to a gallon (3.6 L) of room temperature water and irrigate your grow bags.

YARD WASTE FERTILIZER

Before you turn your yard waste into a liquid plant food, make sure it hasn't been sprayed or treated with pesticides, fungicides, or especially herbicides. The last thing you want is to apply an herbicide-laden fertilizer to your garden.

- Weeds
- Grass clippings
- Water
- Bucket

Collect fresh clippings, weeds, and yard trimmings in a 5-gallon (19 L) bucket and fill it with fresh water. Allow it to steep for at least four weeks to extract as much nutrition from the solid materials as possible. Do this outside, as the rotting materials can start to smell.

When it's ready to apply, pour through a strainer to filter out the solids and apply the liquid mixture to your grow bags.

MANURE FERTILIZERS

If you live in a more rural area or have access to animal manure, this recipe can add an incredible amount of fertility to your garden. Manure is high in nitrogen and this recipe makes excellent use of a true waste product.

- Manure
- Water
- Bucket

If you don't have access to a source of manure, you can purchase granular organic manure products and transform them into a liquid fertilizer. If you're using fresh manure, fill a 5-gallon (19 L) bucket with water and add about a shovelful of manure, then let that sit for a few weeks in an outdoor area.

If you're using granular manure fertilizers, you can steep for less time as these products have already been processed and the nutrients become soluble faster.

A simple prefabricated fence up against two grow bags, growing hops for beer.

Applying Liquid Fertilizers

As mentioned, the benefit of liquid fertilizers is that the nutrients are readily available for your plants. While they're easy to make and apply, there are a few mistakes to avoid:

1. **Using Too Strong a Solution:** Whenever possible, err on the side of diluting your mixtures *more* rather than *less*. You can damage plant roots and turn the tips of your leaves brown by applying too strong a mixture.

2. **Steeping Too Long:** The purpose of steeping is to extract nutrients from solid materials, but there is such a thing as too long. Make sure that you don't forget about your brew.

3. **Improper Application:** Be sure to apply liquid fertilizers to your plants at the base of their roots. Avoid splashing on the undersides of leaves, or even worse, applying directly over the top of your grow bags.

STAKING, TRELLISING, AND SUPPORTING

Don't think you're limited to low-growing bushing plants in your grow bags. The world of trellising, staking, and supporting is available to you as well. Supporting your plants can be more challenging as the sides of grow bags are more flimsy than traditional containers, but there are plenty of creative techniques to help you support needy plants.

Simple Stakes and Clips

The simplest way to achieve support is a freestanding plant stake and mechanism to tie the plant to the support. Bamboo or wooden stakes work well but are recommended only for larger grow bags as they require a certain amount of soil bulk to remain upright, especially when heavy fruiting plants depend on them for support.

For young plants that need time to establish, a small bamboo stake tied to the stem will do.

Large-leaved and shallow-rooted climbing plants, like these cucumbers, need support to grow well.

CROP	SUPPORT
Brussels Sprouts	Minimal staking for support
Bush Bean	Minimal staking for support
Berry, Cane	Staking for support, flat trellis, or circular cage
Corn	Minimal staking until sturdy enough to handle wind
Cucumber	Vertical trellis, horizontal trellis, or circular cage
Eggplant	Staking for support or circular cage
Fruit Trees	Minimal staking for 1st-year trees; none afterward
Grape	T-shaped trellis or metal flat trellis
Hops	Tall vertical wire supports for vining
Kiwi	Long horizontal wire supports for vining
Melon	Vertical trellis, horizontal trellis, or circular cage
Okra	Minimal staking for support or circular cage
Peas	Mesh-covered circular tomato cage, tripod, or vertical trellis
Pepper	Minimal staking for support
Pole Bean	Mesh-covered circular tomato cage, tripod, or vertical trellis
Spinach	None for normal spinach, trellis for everbearing
Tomatillo	Minimal staking for support or circular cage
Tomato	Circular tomato cage, can be reinforced with support stakes
Winter Squash	Vertical trellis, horizontal trellis, or circular cage

Trellising for Different Vegetables

Your trellis or support choice is informed primarily by the type of plant you're growing. Pole beans or peas need 4 feet (1.2 m) or more of vertical support to climb up, but don't need that support to be very sturdy, as their seedpods aren't too heavy.

Cucumbers, on the other hand, need the same or *more* vertical support, and also produce a prolific amount of long, heavy fruits. A more robust support is needed for plants such as these, tomatoes, and so forth.

Bamboo Trellis Designs

If you have access to bamboo, you have access to a wealth of material with which to build. A trellis made of bamboo can last many years if it's built well, and it's often inexpensive or free. It's also extremely hard and able to stand up to some serious weight without breaking.

Let's go over four types of bamboo trellis that you can construct to use with grow bags. Each has its pros and cons, but they're all useful.

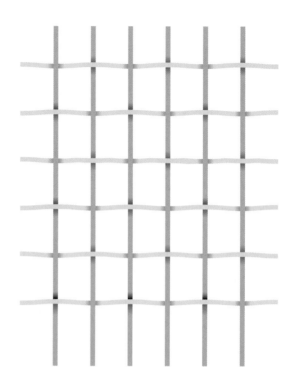

FAN TRELLIS

A fan trellis can be a beautiful thing. This style of trellis is shaped much like a V, wider at the top than it is at the bottom. Usually constructed of three to five long and sturdy bamboo poles with narrower cross-beams, it tucks into the soil at the base of a plant. As a vining plant grows, the widening space allows plenty of room for foliage to spread. The fanlike shape also provides lots of sun exposure and airflow.

Where this fails is its narrow base. Due to its narrower footprint, it has less stability in high wind than other trellis designs might. If it's heavily covered in foliage, the wind can easily blow it over, and that can damage your plant's roots or completely uproot it. It's best used in areas where it's sheltered.

WOVEN TRELLIS

If your bamboo is all of a similar diameter and length, a woven trellis might be great. This method's good for freshly cut and flexible young bamboo. As it dries, it hardens, becoming more rigid.

Woven trellises can be built to personal preferences. For instance, peas or beans don't have lots of weight but do have height, so spacing your vertical and horizontal rails 5 to 6 inches (13 to 15 cm) apart will still support them. Plants with lots of weight may require additional reinforcement, and placing the rails closer together will provide that. A squash or cucumber plant is a good example.

But there are potential downfalls to this style of trellis. If you do not secure each cross-point well (with tightly wrapped twine, cable ties, a nail or screw, or some other method), the rails can still slide downward. A heavy plant can also cause warping over time, throwing off the trellis shape.

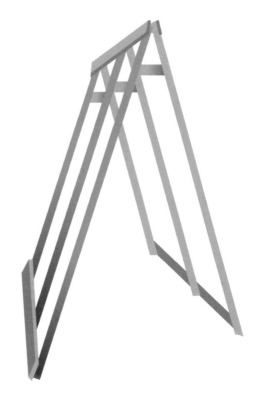

HORIZONTAL TRELLIS

This style of trellis is propped up to the side of your grow bag. It's really useful for cucumbers, squash, or other heavy plants because it supports the fruit off the ground. As the vine grows, it creeps out over the trellis, and the fruit form underneath.

You'll need to make sure it doesn't end up bending in the middle, and that it's secured to the sides of your grow bag. Otherwise, the vine's growth can push it away and cause it to collapse.

TENT TRELLIS

Finally, this trellis uses thicker vertical rails and two separate trellis panels. Leaning these together and using twine to tie the top makes an inverted V, much like a tent. This is extremely effective if your grow bags are underneath, as the foliage shades the base of a plant.

This method is very sturdy in windy conditions but has similar problems to a single-panel woven trellis. In addition, you'll need access to each open end so that you can weed, mulch, feed, or water your grow bags.

PROJECT: A TIPI TRELLIS MADE FROM BAMBOO

There are many different trellis styles that will work very effectively for your grow bags. One of the most popular of these is the tipi trellis. This method has been around for hundreds of years and is very effective.

Let's go over the basics of materials and assembly, and then I'll teach you how to use them in both methods.

MATERIALS AND ASSEMBLY

This method can be extremely cost-effective if you're making a low-budget garden. First, you'll need a selection of sturdy, long, and straight branches. They can be bamboo poles, straight pine branches, young sapling trees, and the like. These should be similar in thickness and length.

You'll also need either a thick garden twine or lightweight rope, a pair of scissors, and perhaps a saw to trim your poles to size.

Before you begin, there's a question you need to answer. Do you want to use one large trellis for multiple grow bags, or do you want to give each bag its own trellis? Some vining plants put out a whole lot of foliage, and they may need a trellis for each bag. Others have one narrow vine with leaves, and those can share a trellis with other bags.

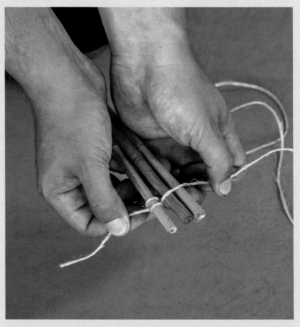

Select the three longest and straightest poles. These will form the main tripod for your trellis and will support the majority of the weight. Lay these side by side on the ground, and make sure they're trimmed to the same height if possible. Make sure that the end you plan to be the "bottom" is perfectly even.

On the other end of the poles, a few inches to 1 foot (about 10 to 30 cm) down depending on length, tie your rope onto one side pole. A clove hitch works very well for this purpose. Once it's tightly secured onto the end pole, wrap the rope around all three poles to tie them together in their side-by-side configuration. Be sure to wrap around all of the posts four to six times.

Once the poles are wrapped together, go around the post that you tied off to and make two loops around your pole-wrapping between that post and the one beside it. Your rope should fully encircle those initial wrappings and go between the posts.

Stand up, placing your weight on the poles briefly to keep them secure, and tightly pull those encircling loops together. Repeat this between the center post and the third post so that the original wrapping cannot unwind. Tie off to the side post but leave a very long piece of rope dangling from the poles. You'll use this later.

Now, set up your tripod. The two outside posts should crisscross, with the center post leaning in between them.

DIFFERENT LAYOUTS FOR SINGLE OR MULTIPLE GROW BAGS

If you're using a single grow bag, you may be able to stop at just a basic tripod. Either set the tripod into the top of the bag and press it lightly into your potting mix or set it up outside the grow bag against the sides. You can wrap any excess dangling rope around your original tripod tie to add extra support and prevent slippage.

Multiple grow bags? No problem. Use longer poles to begin with and set up your grow bags in a horseshoe shape. You'll need more poles for this method depending on the number of grow bags to which you're adding supports.

Take your tripod and set it up so that the first and third poles are sitting inside the two end bags for your horseshoe-shape. The center post should be in the center back bag of the horseshoe shape. Working along the horseshoe from the opening toward the back on one side, set in more poles, leaning them up against the center of your tripod. Repeat on the other side.

Once your poles are all in place, this is when that extra rope is really put to good use. Firmly grasp the rope and walk in circles around the outside of your horseshoe with its tipi frame. You're essentially wrapping over the top of your original wrappings, anchoring all of the new poles in place, and keeping them from coming undone. Once you've wrapped it at least five or six times, tie off the rest of the rope with a half-hitch to one of your original tripod poles.

If necessary, you can add up to two poles per grow bag with this method. Be sure that your original tripod is tall enough that it looks like an inverted tall V, and that your side poles are long enough to lean securely against the original tripod.

Now, install your plants at the base of the poles, help them find their way up the poles, and you'll have a green conical tipi trellis in no time. If you need extra support between the poles, feel free to tie twine between them to create a mesh that your plants can climb.

Growing Next to Existing Supports

The easiest and often sturdiest support systems are those that already exist. Chain-link fences, porches, and even balcony railings can function as a makeshift support in pinch. In fact, one of my favorite ways to cover up an unsightly chain-link fence is to plant a rapidly growing vining plant such as passionfruit, blackberries, or raspberries next to it, and let nature do its cover-up job.

When growing extensive climbing and vining plants in grow bags, sometimes it makes sense to sacrifice some of the inherent mobility grow bags offer for an established, sturdy support structure. For example, when growing luffa gourds in grow bags, I opt to place them next to an existing arbor, as luffa are prolific vining plants that produce dozens of heavy gourds. Building a separate structure contained within the grow bag doesn't make as much sense.

Skip building a trellis altogether if you have some extra paneling lying around that accommodates climbing plants.

PRUNING

Entire books have been written about pruning plants, so here I'll focus on some recommendations specific to grow bags. The most common pruning you'll do in your bags is to remove bottom growth, which has the propensity for disease, rot, or simply inadequate growing conditions.

Removing some growth from a cucumber to open up the interior canopy for more light.

On plants such as cucumbers, tomatoes, or any large bushing or vining plant, eventually the large, bottom leaves may block the interior foliage from accessing light, causing it to start to die. When you see this happen, simply remove the lowest leaves from the plant and toss them into your compost.

PROTECTION

With traditional raised beds, there are all sorts of attachments you can add to provide protection from the cold, heat, and pests. In grow bags and classic containers, these are trickier to set up, but not impossible. With some creative building techniques, we can still achieve an impressive amount of crop protection.

SMALL-SCALE PROTECTION FOR YOUR PLANTS

Maybe building a larger protective frame isn't something you need to do. Perhaps you're just trying to keep the neighborhood cat away from your catmint or keep snails away from your lettuce. What options do you have for plant protection?

Let's go over a short list of different items you can use for grow bag protection.

Cloches: The term "cloche" originates from the French word *cloche*, which literally translates to "bell." This bell- or dome-shaped wire mesh frame is a common way to keep large insects or small rodents away from individual plants. A cloche generally allows airflow to reach a plant and is best for pest-prevention purposes.

Plastic Milk Jugs: If you cut off the very bottom of an empty plastic milk jug, it can act like a mini greenhouse. Pop off the lid (cap) to allow ventilation during warmer weather. This is a great option for someone who's starting plants early in the spring.

Two-Liter Bottles: These work much like plastic milk jugs do but are more translucent, whereas the plastic of a milk jug is slightly milky in color. They're also thicker and can retain more heat.

Tomato Cages: If you have an unused tomato cage around, wrapping it in bird netting, floating row covers, or even cling film can protect the contents of your grow bags. Simply place it over or into the grow bag and then wrap around the wire to keep stuff out.

Laundry Baskets: Do you have an old laundry basket? It can work as a cloche for shorter plants. You may need to add a few holes in the bottom of the basket to allow extra light if it's going to be a long-term solution.

Mini-Hoop Houses: Making a miniature hoop house is surprisingly easy to do. You can use either flexible PVC or green branches and cover either one grow bag or as many as needed.

Wire Mesh Trashcans: The small indoor trashcans that are meant for wastepaper in an office setting can be a very effective cloche over smaller plants. Give them a try.

Tiny Tipi Trellises: You can make a miniature version of a tipi trellis and wrap it in cling film or floating row cover to act as a cloche.

Frost Protection

If you live in a climate where the fall and winter get cold, you might wonder how to protect your perennials. After all, a fabric pot allows a lot more airflow at the root level of plants, which could be a risk. Let's go over some of the options you have available for protecting plants in grow bags from cold conditions.

WHAT FROST/FREEZE CONDITIONS ARE

Are you unsure whether you live in a region where you have annual frosts or freezes? First, it's important to know what they actually are.

In contemporary weather reporting, a "frost condition" occurs when the temperature drops below 32°F (0°C). But for gardeners, there are two levels to be aware of: a "light frost" or "soft freeze" condition occurs when the temperature drops below 32°F (0°C) for just a few hours. In contrast, a "hard frost" or "hard freeze" happens when the temperature drops below 28°F (-2°C) for a longer period.

Most tropical plants should be protected once the weather dips into the high 30°F (-1°C) range. This includes young tropical trees such as mango or banana, but also many common houseplants and a lot of warm-weather annuals. Your tomatoes will be at risk from a light frost and will likely die during hard freeze conditions, for instance.

Some types of plants actually thrive with a little bit of frost. For instance, kale tends to taste sweeter after a light frost. But hard frost/freeze conditions can cause defoliation to most plant types, and in some cases, they can damage roots, stalks, or trunks too.

HOW TO PROTECT YOUR PLANTS FROM FROST

Once you've established which plants are at risk, you'll need to protect them. There's a variety of methods that can be used for each type.

Trees: Saplings that are less than two years old are particularly at risk. These young, often small trees are still in early stages of development.

If it's a deciduous tree, protection for both the trunk and branches may be beneficial. A popular method includes wrapping the tree in fabric or plastic. Bubble wrap or plastic bags are commonly used to loosely cover the lower parts of branches. This can take a little time to do, as well as a lot of tape or twine, but as long as you secure them loosely, the tree will have enough air inside the plastic.

Fabric such as old blanket material works well, too, particularly old wool. Wool stays warm even when it's wet, so if the weather's damp, this may be a good choice for you. Quilt batting is another option that works well, but even inexpensive blankets can help. If all else fails, wrap the trunk in old sheet material.

You can also purchase frost bags designed to go over small trees or shrubs. These are expensive but look nice in the landscape and will protect your saplings.

If they're small enough to be portable, moving a tree indoors for a few days during the worst cold can be helpful. Even if they're inside an unheated garage, they have more protection than they would outdoors. They'll survive for a little while without regular sunlight, particularly if they've gone dormant for the winter.

Remember that there are some types of trees that are used to a cold winter, and in fact, require a certain number of chill hours. These are often okay even if there's snow or ice present. Your real concern should be for trees from regions where frost is uncommon or nonexistent.

Annuals and Perennials: Whether it's a shrub, a crop-producing plant, or even just pretty foliage or flowers, you can provide protection for your smaller plants too.

A greenhouse is, of course, the best option for these. Greenhouses provide protection from wind, and they're relatively easy to keep above freezing temperatures. Even just adding some string lights around the plants inside a greenhouse can increase the relative warmth inside, enough to keep frost at bay.

If you don't have space or money for a greenhouse, don't panic. You can build cold frames out of a variety of materials. Whether they're rigid or flexible, these are usually coverings that allow light through but reduce the amount of cold air directly around your plants. A simple cold frame could be as easy as clear plastic over a PVC framework, and it doesn't cost much. More elaborate ones can use old window panels or clear plastic sheeting to make a rigid structure.

Just as with trees, you can use old bedsheets to wrap larger plants. Be sure to leave room when possible so that their leaves have air space.

Floating row covers are a godsend. Made in a variety of thicknesses, these meshlike fabrics can block most of the cold really well. They also keep pests at bay, which is particularly useful in the very early spring when plants are just starting to put out new growth.

GROW BAG–SPECIFIC TECHNIQUES

Clustering a number of grow bags close together can help keep the soil warmer. This is especially important if you have any tropicals, as you can place them in the center of a ring of other plants. It also makes it easier to cover a larger number all at once.

If you are concerned about moisture control and airflow around the foliage, you can tuck straw or crumpled newspaper between fabric pots. This provides the same protection as placing fabric pots against one another but allows more space between individual plants.

Long stakes or sections of PVC pipe can be placed at three or four points at the sides of a grow bag to act as supports for floating row cover fabric or plastic sheeting. Use some twine to tie it in place around the bag but be sure that you can easily remove these when the weather warms again.

Plan your overwintering location in an area where light still can reach, but that's protected on the sides from cold wind. This will reduce the amount of "advective frost," a type of frost caused by wind chill. It doesn't prevent "radiation frost," which is frost from still air with extremely cold conditions, but your other methods of protection guard against that type of frost.

PROJECT: MAKING A COLD FRAME FOR YOUR GROW BAG

As the summer slips into the fall, or even in very early parts of spring, a cold frame may be a real benefit. But how do you make this work when you're using a grow bag?

A frost blanket works extremely well to replace a more rigid cold frame. It still needs support to keep it off the foliage, but it will increase the temperature around the plants and protect against the chilly air.

Let's go over what you'll need to make a frost blanket cold frame for your grow bag.

MATERIALS

For the simplest and least expensive version, you'll need four or five sturdy stakes, a frost blanket or some floating row cover fabric, and some twine or other means to secure it. This is usually an emergency method intended as a temporary solution until you can build something more resilient.

Looking for something that's inexpensive but a little sturdier? Get PVC pipe, eight pipe corners that the pipe fits perfectly, some pipe sealer, a frost blanket or floating
row cover, and some wire or twine to secure the material.

For a much sturdier version that'll hold up to serious weather or wind, you'll need 2x2 (5 x 5 cm) lumber, a saw, nails, a hammer or a nail gun, a frost blanket/floating row cover fabric/heavy-duty plastic, and a heavy-duty stapler.

A TEMPORARY FROST COVER

Not being prepared for a cold snap happens to all of us. But if you keep some stakes and floating row cover on hand, you can assemble an emergency cover for your plants.

Slide the stakes in along the sides of the grow bag, evenly spaced; about four or five should give you plenty of support for the fabric. If you're using a very large grow bag, you may need more. Make sure your stakes are pushed down to the bottom of the bag to provide enough support.

Once the stakes are in place, put your frost blanket/floating row cover fabric over the plant. Make sure it's covered on all sides. Then take twine and tie the fabric around the stakes to hold it in place; about two or three tie-points should keep it from moving. Add one more around the grow bag itself.

This will work for a day or two, but you'll need something better for longer protection.

PVC FROST COVER FRAME

Begin by measuring across the widest point of your grow bag. You are going to make two squares out of PVC pipe and corners that can easily slide over the widest point of your grow bag. The length you'll cut the eight pieces of PVC for the sides of those squares will be just slightly wider than your grow bag.

Then figure out how tall of a frame you'll need to cover the grow bag and the plant entirely. Cut four lengths of PVC at that height, making sure they're all an even length. Use PVC corners to assemble the two squares, then add your height posts between the two squares to make a long rectangle.
Place this over your plant and drape the frost blanket/floating row cover over the top. You can use wire to go through the cover and around the PVC poles. Alternately, use twine to tie the fabric at multiple points along the frame to keep it in place.

HEAVY-DUTY FROST COVER FRAME

Are you expecting some high wind that would topple your PVC? You can do the same shape of frame out of wood. If you're expecting really severe conditions, leave off the bottom square shape of the frame and make the side posts longer, and use them like stakes into the ground around your grow bag.

Since you're constructing this of wood, you can skip the twine or wire and use staples to tack the material directly to the frame. This method also works well when using greenhouse plastic sheeting, but you may need to provide vent holes at the top to allow excess heat to escape during the day.

Heat Protection

Aptly named, shade cloth helps cut down on heat and blinding sun, relieving plants of those stressors. Plants have an upper limit of how much heat and sunlight they can tolerate, and if exceeded, can suffer from sunburn and heat stress.

Using shade cloth will impact the growth and appearance of your plants. If you grow under shade cloth for long periods, many plants respond by increasing their leaf size, growing taller, and other adaptations. This is plant-dependent, but the general idea is that all changes you make to the environment in which your plants are growing will influence their growth.

The two considerations to know when purchasing shade cloth are the material and the density percentage.

The density of shade cloth you choose is the most important factor. It's a percentage rating that tells you how much sun is blocked by the cloth. Here are some general ideas of what plants like for various percentages of density shade cloth.

DENSITY	RECOMMENDED PLANTS
30%	Pepper, tomato, eggplant, squash
50%	Flowering plants
60%	Lettuce, kale, spinach
70%+	Ferns, alocasia, other houseplants

SHADE CLOTH: HOW DOES IT AFFECT GROWTH?

People in hotter climates often find shade cloth to be useful in protecting their plants from the scorching heat. But does color actually matter for your plants?

Answering that question takes a little time. Before we can talk color, we need to talk about the cloth itself, along with the percentage of shade that it provides.

The density of your fabric itself plays a major role in how much light reaches your plants. The material itself can be knitted or woven, but the spacing of the knit or weave determines the amount of light that can filter through.

Different percentages are used for different plant types. Let's go over the most common:

- **30% Shade:** This is often used for heat-tolerant plants that just need a tiny bit of sun protection in the midst of summer. Tomatoes, peppers, both summer and winter squash, and other heat-tolerant plants are best for this. Some heat-tolerant flowering plants, such as geraniums or snap-dragons, may also do well at this percentage.
- **40% to 50% Shade:** This half-strength range of lighting is good for nearly all flowering plant types. It's especially beneficial for bulbing plants like lilies or camellias but can be used with great effect for more light-tolerant orchids.
- **60% Shade:** Growing your greens throughout the summer? This level of shade may be of use. The sun's rays do still manage to get through, but it provides a much greater level of protection for your lettuce, spinach, or bok choy.
- **70 to 90% Shade:** Fern-lovers rejoice, this shade cloth is made for you. Simulating the effect of a full tree canopy, this range provides lots of protection for understory plants like ferns, philodendrons, and dracaenas. This level is also most commonly used for shading humans, so it's widely available.

KNITTED OR WOVEN?

Is there a real difference between knitted fabrics and woven ones, or is it all cosmetic?

Knitted shade cloth typically is constructed of lightweight polyethylene, or PET. Because of how it's constructed, it resists tears or fraying, but it can get snagged on thorns. More widely spaced than woven fabrics, it allows great airflow. This makes it a little more resistant to wind damage and allows more heat to escape from under the fabric.

Most knitted shade cloth is UV-resistant, and it's often used for greenhouse fabric. But it's not without a fault. This fabric can stretch or shrink over time by as much as 2 to 3 percent depending on the weather conditions it's experiencing.

Woven shade cloth is made of polypropylene, or PP. It's UV-stabilized to hold up against even the most significant heat exposure and doesn't typically stretch or shrink. Its spacing is much closer, making it the perfect choice for patio shading or privacy screens. It also provides a lot of sun protection for plants.

However, it's heavier in terms of weight, so you'll need a much sturdier frame. Because of how it's woven, it can fray at the edges and needs to be hemmed. And that closer spacing also allows less heat to escape from underneath the cloth.

PROJECT: PROTECTING MULTIPLE GROW BAGS FROM THE ELEMENTS

Are you looking to put up a floating row cover to protect your plants from pests? Perhaps you're considering a fall crop and want a little protection from the chill, or maybe you're looking to provide shade cloth to ease the summer heat.

Even something as simple as staking up a frost blanket around the perimeter of your bag can protect plants well.

No matter what the purpose, a remarkably simple frame can provide a cover for an entire row of grow bags. This project is made of PVC piping, PVC corners, PVC four-way connectors, your cover of choice, and, strangely enough, duct tape. And it works surprisingly well.

Begin by figuring out how big of a frame you'll need for your grow bags. Line them up and use a measuring tape to plot out the length of your row. If it's particularly long, leave a 2- to 3-inch (5- to 7.5 cm) gap every 3 to 4 feet (0.9 to 1.2 m) between the bags. You'll want the frame to extend out 6 to12 inches (15 to 30 cm) beyond each end of the line of bags, so add on 12 to 24 inches (30 to 60 cm) to the length of the bags themselves.

Once you know your length, divide that up into 3- to 4-foot (0.9- to 1.2 m) segments. If, for instance, you're doing a 12-foot (3.6 m) row of bags, you can split it into three equal segments of 4 feet (1.2 m) or four equal segments of 3 feet (0.9 m). At each of those segment points, you're going to have a curved support pipe, so if you're using heavy fabric, go for shorter gaps.

Now, take your PVC piping and cut pipes to the length of each segment. Using the example of a 12-foot (3.6 m) row, you'd need a total of eight 3-foot (0.9 m) lengths of piping. Set those aside.

Once you've got your sides for the length, it's time to determine the height and width of your frame. How tall do you want your cover to be? You'll use one PVC pipe bent into a curve to create this height. As a general rule, a 4-foot-tall (1.2 m) cover will need at least 8 feet (2.4 m) of pipes. A 2-foot-tall (60 cm) cover will need at least 4 feet (1.2 m) of pipes. The width is the same as your planned height.

Using the same example of a 12-foot (3.6 m) row, assume we want a 4-foot-tall (1.2 m) cover that can protect a row of peppers. We would cut four 4-foot-long (1.2 m) posts to create the width of the container, and four 8-foot-long (2.4 m) posts to create the frame. This means we now have eight 3-foot (0.9 m) pipes, four 4-foot (1.2 m) pipes, and four 8-foot (2.5 m) pipes.

Now it's time to build your frame. Use three-way PVC corners at the four outermost corners. All other joints will need four-way PVC joints. Make a series of 3 feet (0.9 m) long by 4 feet (1.2 m) wide squares side by side. Use the connectors to join them all together into one long, continuous frame around your grow bags.

Take your longest poles and go along the long side of the frame, inserting it into the remaining hole on each connector. Make sure all your connectors are tightly attached and won't come undone, then bend each long pole into a hoop-house shape and insert it into the connector on the other side.

If you want to add extra reinforcement, use one leftover four-way connector and two 6-foot (1.8 m) pieces of PVC to make a top rail. This will prevent the cloth from sagging down in between the curved segments. Use the duct tape to secure this pole to the center of the top of each curved pipe. Duct tape is great for this as it won't snag your fabric and cause damage like a screw might.

Once that's secured, your frame is complete, and all you need to do is drape your fabric of choice over the top. This half-circle or half-oval long frame can be left in place year-round if you want to swap between cold protection and heat protection or pest prevention.

So now that you've got a good background on the different types of shade cloth, what's the difference between a knitted 50 percent black cloth and a knitted 50 percent white cloth? Or is there any difference at all?

The answer is a definitive *yes*. While studies are still being done on some less-common colors of shade cloth, such as blues or browns, the most common shades are well known in terms of how they function.

A lighter color shade cloth, such as a white, tan, or yellow, can reflect the sun's rays more effectively. This means that it'll be cooler underneath the cloth. This can be really important if your shade cloth is woven, but it has benefits for knitted cloth as well.

But light-colored shade cloth also tends to allow a lot more light through. This is fine if you're growing plants such as tomatoes but may not work as well on your full-shade-loving plants.

Dark fabrics such as green or black often allow less light to pass through, providing a more complete shade. But these darker fabrics can also allow more heat through. Knitted shade cloth might be better if you're concerned about heat buildup underneath these. Black is the worst for heat buildup, whereas green is a little less warm.

Do colors have any other uses? Surprisingly, yes.

Apparently, orchids seem to really like growing under blue shade cloth. They produce more foliage underneath blue fabric.

Lettuce appears to love the color red. Under a red cloth, lettuce stems and leaf development are larger. This works for philodendrons as well, which produce

larger and healthier leaves underneath red fabric. But philodendrons don't like blue at all, producing smaller leaves when placed underneath a blue fabric.

Peach trees love shade cloth no matter what the color is, but they don't need as high of a percentage cloth under light colors. A 10 to 15 percent white shade cloth is equivalent to a 30 percent shade for grays, blues, reds, or yellows. In both cases, they produce denser foliage.

No matter which color you choose, shade cloth is an effective sunscreen for your plants. And on those scorching summer days, we could all use a little shade.

Pest Protection

With all gardens, there's always the risk of pests, diseases, or even the neighbor's cat wreaking havoc. Grow bags are no different in that regard. But what is different is the methodology that you'll use to deal with pests in your grow bag gardens.

An integrated pest management (IPM) system is usually the best way to deal with the majority of your pest problems. While this sounds complex, it's actually one of the easiest ways to ensure your plants are in perfect health and safe from external damage. Four basic tenets handle the vast majority of pests and diseases.

Whether you're trying to prevent cabbage moths or cucumber beetles, reduce root rot or powdery mildew, the four sections below should help you. Each segment takes one of the facets of IPM and explains how it works. Combine the four and you'll have a lush and verdant garden.

BETTING ON BIOLOGICALS

Most pest insects have several predatorial insects that would love to eat them. By introducing beneficial species of insects to your garden, you can control quite a few different types of pests.

Flying pests such as moths tend to leave eggs on the underside of leaves. These eggs are often preyed upon by ladybugs, lacewings, or other species. Providing housing for your beneficial insects can entice them into living in your garden year-round and producing generations of garden helpers to come.

Predatory wasps are tiny insects, quite different from the large and scary-looking wasps we're all familiar with. These tiny wasps lay their eggs inside of moth larvae, and as the eggs hatch, their young eat the larvae from the inside out. They also feed upon some forms of beetles in a similar way.

Praying mantises can be a godsend to a garden too. These large insects not only eat a lot of pest insects, but they're also fun to watch as they're quite territorial and will defend their chosen plants.

What about soilborne pests? Grubs and root knot nematodes will fall prey to a few different species of beneficial nematodes. These microscopic round-worms won't hurt you or your pets, and in fact, you'll never see them. But as long as the soil is moist they'll dwell within and protect your plants with surprising success.

Don't forget that you have other biological methods at your fingertips too. Some forms of mycorrhizal fungi and some bacteria are extremely effective as control methods for pests. For instance, *Bacillus thuringiensis* is a bacteria that will happily consume pest insects. There are different species of this bacteria, but one of the most common is *Bacillus thuringiensis* var. *israelensis*. That subspecies is incredibly effective against most caterpillars and is sold as an all-natural caterpillar-killing product.

Certain mycorrhizal fungi form a symbiotic arrangement with the roots of your plant. They provide nutrients to your plants, and in return share the plant's space and water supply. Some of these species can also provide some protection against pests, particularly types that attack a plant's root system.

Whenever possible, provide an environment that encourages these biological control methods. As long as they have a safe place to live and an ample supply of food, you'll have half your work done by the creatures themselves.

CREATING A CULTURED GARDEN

Cultural controls are methods used to create the right conditions to avoid disease and pest infestation in the garden. These methods range from proper watering technique to pruning and plant maintenance.

For most grow bag gardeners, there are two cultural controls that are essential: pruning to allow for good airflow and drip irrigation to reduce water splash back onto plants. Since soil can splash up onto the leaves of plants, it can carry pest eggs or plant diseases within it. Pests already present on a plant have easier access to the rest of the plant if it's growing too tightly together.

Fast-growing plants should be pruned regularly to ensure there's good spacing surrounding the plants that allow air to pass by. Diseased fruit should be removed quickly, and in the case of trees such as apples, removing all but one from a tight cluster will reduce the spread of fruit pests and make the remaining one on the tree much more flavorful.

Mulching around the base of your plants reduces watering frequency and prevents soil splash back, so it's another excellent example of a cultural control technique.

BUILDING BARRIERS

Certain types of physical barriers are also fantastic controls for pests. Some mechanical devices also fall into this category.

If you're contending with rodents, a trap is an example of a mechanical control method. Traps can be useful for most mammals, and there are both safe catch-and-release methods or damaging traps. If doing live capture, you'll need to find an appropriate place to release your caught animals, but this can prevent you from losing your harvest to wild rabbits, squirrels, or even skunks.

For most of the pests that gardeners encounter, a physical control such as a floating row cover can keep smaller insect pests at bay. It doesn't work against pests that crawl, such as ants, but it is extremely effective against flying pests such as moths or some types of beetle. Use a framework to place a floating row cover over either a single grow bag or multiple bags all at once, and you can keep the flying bugs out.

Remember that if you've got plants that flower and then set fruit, physical barriers may need to be removed when the plant begins to flower. Otherwise, pollinating insects won't be able to reach the flowers. If there are a lot of pests in the area when your plants flower, you can hand-pollinate using a fine-tipped paintbrush or a cotton swab, thus allowing you to leave the floating row cover in place.

Mulches also technically fall into the category of a physical control method as they reduce the spread of weeds. As weeds can choke out your other plants and also can be a home for flying pests, a mulch is well worth your effort!

Diseased soils, such as ones infested with fusarium fungi, can be steamed to kill plant disease causes. Other methods include soil solarization, which is done by superheating the soil by placing something black such as plastic on the soil's surface for a long period. Controls like this will kill weed seeds, fungi, and bacteria in the soil, but do not allow good plant growth. Still, a technique such as this can be put into place in an otherwise fallow planting area when you're working another location.

Finally, fencing can be used to keep larger pests out. This prevents deer from nibbling your plants away but may not be as effective against digging pests.

TREATMENT WHEN ALL ELSE HAS FAILED

A combination of these things will reduce pest and disease issues immensely. But flying insects are a risk at any time, as they're mobile and able to appear suddenly. So, if all else has failed, it's time to consider sprays or other methods of control.

There are many organic pesticides on the market nowadays, many of which are OMRI-rated. If you'd prefer to avoid the chemical methods, those are always available to you. A perfect example is azadirachtin, a natural compound found in the neem seed (usually sold as neem oil). Another is pyrethrin, a pesticide derived from the chrysanthemum flower. A third, spinosad, is a compound normally found in soil bacteria that's effective against smaller, soft-bodied insects such as aphids and fruit flies.

If a combination of the methods listed don't handle all your pests, you can use limited spot treatment of these organic sprays when you absolutely need to. This reduces any concerns about eating your produce later, as it may never actually be treated on that specific plant. In addition, most of these organic methods break down quickly in the environment and are less likely to pollute the water table.

Limiting their use also ensures they're more effective in the long term. Constant use of any pesticide can cause local pest populations to build up an immunity to it, at which point it's no longer useful at all. As just one facet of an integrated pest management system, these sprays can be the option when nothing else works, and the end to a particular pest woe when it arises.

PUTTING IPM TO USE IN THE GROW BAG GARDEN
So now that you have more of a grounding in integrated pest management, how does it actually all apply to grow bags?

If you're like most people who've adopted this gardening method, you have more than one grow bag. You can group bags together based on height and create a floating row cover that protects them all as a physical control method.

At the same time, you can lay out custom drip hose across the bags to provide water at the soil level, thus reducing your need to water from overhead. Mulching the bags will protect them from multiple problems. Adding a root zone fungal blend or beneficial bacteria to the soil in the grow bags can boost the health of your plants, and at the same time you can apply some beneficial nematodes to take out soilborne pests.

Place a ladybug house near your grow bag garden, and these cheery little red bugs will take out many pests that are outside your gardening zone. If the pests never make it to your fabric pots in the first place, they can't harm your plants. A single praying mantis cocoon placed under a floating row cover will hatch and provide you with another method of protecting those plants as well.

Maintain good pruning for airflow and disease prevention on your plants and keep a watchful eye on them. Lift the cover and inspect your plants regularly for signs of damage and handle it when it appears. You may never have to resort to pesticides, but they're available just in case they're needed.

Mold or algae growth is natural when you have water, nutrients, and light.

PROLONGING THE LIFE OF YOUR GROW BAGS

While it's easy to replace your grow bags, what if you'd like to get more than one season out of them? If cared for properly, a grow bag can last for many growing seasons, and they're easy to store when not in use.

Managing Mold and Stains

It's common to see a whitish or greenish mildew-like substance on the bottom part of your fabric pots. Whether it's mold, hard water scale, or even salt buildup on the outsides of your pots, it can be pretty unsightly.

While your grow bag is full of soil mix, prepare a soft-bristled brush and a bucket of baking soda dissolved in water. Dip the brush in the baking soda, and then lightly scrub the surface of the pot. Most of the mold or stain should come off.

Elevating a grow bag can also help reduce mold or mildew. A constantly damp bag that's exposed to a wet surface more readily molds. By elevating it, you allow airflow around the bottom of the bag, reducing the likelihood of mildew buildup.

Clean Your Grow Bags

It's easy to hose out or sterilize a plastic pot by wiping it out with bleach. But what about a fabric pot?

When cleaning your own grow bags, choose tools that appropriate for the size of the job.

By "hiding" the bottom of the bag between barriers, you lose less water to evaporation while still allowing air pruning.

When winter comes, simply clean, fold, and store your grow bags for the next season.

If you have a washing machine, your problem may be solved. Begin by rinsing the pot to remove as much dirt as you can, then let it sit for twenty-four hours in a dry place. Use your hand or a dry brush to brush off as much remaining soil as you can.

Then pop it into the washing machine on a cold-water cycle. Use either a blend of baking soda and vinegar as your detergent, or an organic detergent that's gentle on the fabric. You'll want to use the "heavy soil" setting, but it's essential that you wash in cold water to prevent severe stretching or warping. When the cycle's done, hang up your fabric pot somewhere to air-dry completely.

Avoid adding bleach to a washing cycle for a fabric grow bag. Nonchlorinated detergents are best if you're not using baking soda and vinegar as your detergent.

Shield Bags from the Elements
Over time, sunlight beating down on your garden does horrible things to a wide variety of materials. Your grow bags are at risk too. Similarly, they can be at risk from excess moisture, from freezing conditions, and more.

Shielding your bags from the elements can help prolong their longevity. If you position your fabric pots in a place where the sun reaches the plant but not the bag, they're a bit more protected. Elevating

them to keep them out of puddles reduces damage from moisture exposure. Bringing them indoors in the winter prevents them from freezing solid and cracking.

Placing your grow bags inside slatted-style containers such as milk crates, old damaged plastic pots with holes in them, or even large paper bags can help protect the fabric of the bag. Anything that reduces direct exposure to the elements can improve their longevity.

Store Grow Bags Properly
When they're not full of soil, keep your grow bags securely stored away. But to do so, you'll have to clean them out and let them completely dry first.

If there is any moisture in the fabric of your grow bag, that's a potential mildew point during storage and can create a weak spot in the fabric over time. Once dry, they're safe to store.

But what's the best way to store them? I recommend keeping them in a dark, secure location. You can put them inside a cardboard box or set them on a garage shelf. Try to keep them in a location where rodents won't find them and shred them to line their nests.

A variety of herbs and flowers in bags with folded lips to add stability.

CHAPTER 6

GROW BAG PLANTING IDEAS

The beauty of grow bags is their adaptability to whatever you want to grow in your garden. Fruits, vegetables, trees, shrubs, flowers . . . the only limitation to your success is your own imagination. In this chapter, I'll share some creative planting ideas that you might not have otherwise considered to show you how far you can take this growing method.

From microgreens to stir-fry gardens, I'm always finding new ways to combine plants to achieve a goal, whether that is to grow a "meal in a bag" or to create a portable pollinator attractant that I can place near my squash, cucumbers, and fruit trees.

Remember, the following combinations are only a starting point. You can adapt these "grow bag templates" to your own favorite styles of cuisine, floral arrangements, fruits, and more.

Multi-pocket grow bags are perfect for kitchen gardens, as you can plant six or more different veggies in the same bag.

KITCHEN GARDEN COMBO BAG

This planting combination requires a multi-pocket grow bag, which you can buy online or make on your own by cutting horizontal slits in a large grow bag. The inspiration behind this bag is the classic kitchen garden, a garden space typically located close to the back door of a home and planted with the most common kitchen herbs and veggies.

With this combination, even the most space-challenged gardener can have a miniature kitchen garden in a grow bag. I like planting bushier leaf lettuce varieties at the top of the bag to provide extra shade cover to the soil surface. This helps keep the soil moist longer, meaning you'll have to water it less often. You can also grow mint over the top of the grow bag, and plant other greens such as Swiss chard, bok choy, and so forth to achieve the same effect.

TOP OF THE BAG
- Leaf lettuce
- Chard
- Mint
- Bok Choy

BAG POCKETS
- Basil
- Parsley
- Oregano
- Thyme
- Sage
- Chives

Pick your six favorite herbs and plant those in the side pockets of the grow bag. Herbs such as oregano and thyme tend to creep along soil surfaces, so isolating them to their own pockets is a great way to get a solid yield of herbs without overwhelming the bag.

On top, plant a mixture of your favorite leafy or Asian greens. Loose leaf lettuce is a fantastic option as its sprawling leaves help protect the surface of the soil on top of the grow bag and are well suited to cut-and-come-again harvesting.

STRAWBERRY BAG

If you're apprehensive about entering the world of growing fruits, growing strawberries is the perfect place to start. They're much easier to grow than fruit trees, blueberries, or other bushing fruit varieties. They pay you back in the same season with sweet, delicious fruit and even propagate themselves via runners.

The compact habit of most strawberries makes them the perfect choice for a multipocket grow bag. There are a few different types of strawberries that exist, and within those types, multiple varieties. It's no wonder gardeners get confused when deciding which to grow.

June-bearing: As its name implies, these types produce all of their berries in a two- to four-week window somewhere around the month of June. They also tend to produce the largest, juiciest strawberries, making them an extremely popular pick for home gardeners as well as commercial growers.

These berries have subcategories of early, middle, and late season. A great strategy for growing June-bearing strawberries is to divide the number of plants you want to grow by three, and select an early-, mid-, and late-season variety of June-bearing strawberries. Doing this stretches your harvesting window from about two weeks to around four weeks. It's not much more, but it's better than nothing.

Everbearing: Despite its name, everbearing strawberries don't produce all year long. You can typically harvest them once in the spring and once in the fall. Only in the most perfect growing conditions do everbearing varieties produce three times in a single season.

These varieties don't spread out as much and spend a lot of their energy producing the next harvest of strawberries. This makes them perfect for vertical systems, or, as you may have guessed, a multi-pocket grow bag.

Day-neutral: These unique varieties will produce strawberries year-round, as long as the temperature is within 35°F to 85°F (1.7°C to 29°C). The downside is the fruit's size, as they're quite a bit smaller than the massive June-bearing varieties. However, if you'd rather have consistent production of smaller berries over a longer period, they're a fantastic choice.

BARE-ROOT OR POTTED STRAWBERRIES?

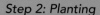

If you're not growing your strawberries from seeds, you have two choices on how to buy strawberry plants: bare-root or already potted up. Potted plants, as the name implies, are already properly planted, growing, and will establish themselves much faster in a grow bag than a bare-root strawberry.

However, they're quite expensive compared to bare-root strawberry plants. You can get a couple dozen bare-root plants for the same price as a few potted plants. If cost is a concern, opt for bare-root strawberries. Planting bare-root strawberries requires a bit more finesse, as there is no soil to indicate where the crown begins and roots end. The biggest mistake you'll make is not planting the crown at the proper depth, so if you're worried about this then you may wish to opt for potted plants that you can simply transplant into your grow bag.

Step 1: Grow Bag Setup
Fill a multipocket grow bag with some moistened Epic Grow Bag Mix (see page 79) to the brim and tap the bag on the ground to allow the soil to settle. Fill with more soil and tap again; you want to remove any large, loose air pockets.

Step 2: Planting
Select the strawberry variety you'd like to plant and source your plants. Growing from seed, while possible, takes much more time and can be fraught with frustration due to how delicate strawberry seeds are.

Purchase plants with big crowns and healthy roots. Be extra careful to plant each strawberry plant at the depth where its crown is perfectly flush with the surface of the soil in each pocket. Planting shallowly will cause its roots to dry out (and the plant will die), and planting too deeply puts the crown at risk for stunted growth or rot.

Firm up the soil around the roots and water well.

Step 3: Ongoing Care
Depending on your level of patience, you can prune your plants in the first year of growth to encourage abundant production in the years to come. Removing flower buds in the first year allows plants to establish their roots and shoots to provide ample energy for huge yields in subsequent years.

Keep your grow bag moist using strategies from chapter 5, as strawberries don't tolerate dry soil well. Fertilize with a balanced organic fertilizer before planting, as well as after fruiting. Avoid fertilizing immediately prior to the fruiting season as this can impact fruit quality and size.

POLLINATOR GROW BAG

As I've progressed along my gardening journey, my respect for the importance of pollinators has reached mythic proportions. Whereas in the past I held on to a mantra of, "If I can't eat it, I won't grow it," these days I plant my gardens with ample numbers of pollinator plants sprinkled throughout.

Grow bags offer a unique pollinator benefit, as you can place them near veggies and fruits that benefit from a robust natural pollinator ecosystem. Instead of running around your garden with a toothbrush, hand-pollinating all your cucurbits (squash, zucchini, pumpkins, etc.), you can plop a pollinator grow bag right next to your squash patch and let nature do the work.

First, consider what's the right pollinator for your crops. That's what you'll want to attract. Let's stick with the concept of pollinating your squash patch.

Most squash blossoms are large, yellow, and somewhat fluted in shape. They're big enough that bees can happily crawl around inside and, in fact, even some types of wasp would fit. Once they find the flowers, you'll be set.

So, you'll want to include plants to encourage more bees and wasps to stop by. Bees love flowers in purple or blue, and wasps are attracted to yellows and browns. Both will explore every flower they find, including the squash you want them to visit, but your pollinator bag will be the reason they come by in the first place.

The selection of plants you choose should share a love of the same basic conditions. There's no point in mixing up moisture-loving plants with ones that prefer it on the drier side. One type won't be happy and will suffer. Pair plants that share similar care requirements, if possible.

Marigolds remain one of my favorite pollinator and pest-prevention combination flowers to plant in my bags.

Let's pair of a selection of drought-tolerant perennials that will entice even the most stubborn of bees to swing by:

- **Pink Creeping Thyme** (*Thymus praecox* ssp. *arcticus* 'Coccineus')
- **Yellow Pineleaf Penstemon** (*Penstemon pinifolius* 'Mersea Yellow')
- **Walker's Low Catmint** (*Nepeta racemosa* 'Walker's Low')
- **Pastor's Pride Lavender** (*Lavandula angustifolia* 'Pastor's Pride')

Here we have four plants that can be paired together in a 25-gallon (95 L) grow bag to build a pollinator garden that will entice in bees from spring well into the fall. Since there are four different heights of plants, you'll want to layer or stagger the plants in the bag so they all receive light. All four of these species are somewhat drought-tolerant and all are bee-friendly.

At the front of the bag where the sun can reach, plant at least two pink creeping thyme. Doing so will ensure that the thyme starts to drape over the side of the bag as well as creep around the other plants. It produces tiny pinkish-purple flowers across its foliage while it's in bloom, making it easy for a bee to land on the stem and visit multiple flowers while there. This will often flower from mid-spring throughout the summer.

Next, we have yellow pineleaf penstemon. This plant usually stays in the 12-inch (30 cm) tall range and it produces huge quantities of yellow tubular flowers all through the summer months. Some of the front foliage will hang down over your thyme.

Behind that, plant Walker's Low catmint. Reaching an average height of 18 inches (46 cm), it unveils mid-blue flower spikes throughout the spring. This will lure in pollinators early in the season, right when the bees are searching for their favorite places to visit.

And finally, we have Pastor's Pride lavender. This tall 24-inch (61 cm) variety has fragrant purple flowers. It blooms initially in late spring to early summer, but if you deadhead the flower spikes once they're spent, it'll put up a second showing of flowers in late August to early September.

As the taller flowers die back, trim them out to refocus the plant's energy on making more. Leave the thyme to its own devices, as it is mostly there as a living mulch and ground cover. These plants all like well-draining soils and don't need to be fertilized regularly, although an annual application of compost around their bases will rejuvenate them.

This bag will come back year after year, keeping you awash in bright color from mid- to late spring all the way into fall. In the winter, make sure it's protected from winter's chill with a cold frame.

BLUEBERRY GROW BAG

One of the most difficult issues for new blueberry growers is that they're acid-loving plants. Many gardeners work really hard to neutralize their soil so that it's great for other vegetables, but for blueberries and some other berry types, that is more of a problem than a benefit.

Make sure to select a blueberry variety with the right "chill hour" requirement for your climate.

The easy solution, of course, is container growing. When growing in a grow bag or other container, it's easy to ensure that each plant has the soil type and soil pH it prefers. And blueberries are no exception to this rule.

What Blueberries Grow Best in Grow Bags?

Technically speaking, you can grow any blueberry variety in a grow bag if your bag is the right size. There are two major types of blueberries: highbush and lowbush. Highbush blueberries (*Vaccinium corymbosum*) are most commonly grown in Mediterranean climates, where the weather is warm and the winters are mild. Lowbush blueberries (*Vaccinium angustifolium*) are most common in colder climates.

Highbush blueberries range from 4 to 8 feet (1.2 to 2.4 m) tall when they're fully grown, and they can be a pretty large-size bush. These will need at least a 45-gallon (170 L) grow bag by the time they're fully grown, although for the first year or two you can gradually increase the size.

Lowbush blueberries tend to stay under 4 feet (1.2 m) in height but have a more sprawling form. While they'll benefit from a larger bag, they don't need quite as much space as the highbush types do. With these, it's usually fine to use a 20-gallon (76 L) grow bag, and you can always size up from there if needed.

A few blueberry hybrids are optimized for container growing. These can grow in a space as small as a 10-gallon (38 L) grow bag and are great for small-scale gardens. These hybrids are usually referred to as "half-high" blueberries.

Whichever variety you decide to grow, try to have at least two plants of the same type of berry. Some types of blueberry need a second pollinator plant, and highbush and lowbush don't cross-pollinate well. There are some self-fruitful types, but even they produce a larger harvest with other plants nearby.

Planting Your Blueberry Plants

Begin by preparing your soil. Blueberries require soil that's between 4.5 and 5.5 pH, much more acidic than most plants prefer. While the lowbush types can tolerate more alkaline soil, they'll produce a better-quality harvest when they have that soil acidity to draw on.

A good acidic soil blend for blueberries is a mix of sphagnum peat moss and pine bark mulch. Aim for at least 40 percent peat, although you may be able to go as high as 50 percent. This will break down to form a nice acidic medium for your blueberries to grow in.

Pick a sunny but sheltered spot. Your berries need protection from the wind. Plant at the same depth it was in its original pot. If it was a bare-root plant, make a mound in the center of the grow bag and spread the roots out over the mound. Be sure you don't cover any graft joints when planting, as those need to be exposed to the air.

Mulch on top to prevent moisture evaporation. Leave a gap between the trunk and the mulch, as this allows the proper amount of airflow.

Blueberry Bag Care and Maintenance

Now that your plants are in place, you'll need to keep the soil consistently moist. Blueberries love to have ready access to the water they need. At the same time, it has to drain off well. Your grow bag and the soil medium should help significantly with this. Aim for 1 to 2 inches (2.5 to 5 cm) of water per week as a goal range, with more provided during hotter weather.

For the first year or two, pinch back flowers as they form. This redirects the plant's energy into growth rather than fruiting. By doing this, you'll have a much healthier plant in the long term.

In early spring before any new leaves emerge, fertilize with an acidic fertilizer meant for blueberries or other acid-loving plants. Fertilize again in late spring if the plant is still young, but mature plants shouldn't need more than a single annual feeding.

Skip pruning for the first three to four years. After that, highbush pruning is an annual late-winter chore so that the plants are ready for new foliage growth in the spring.

Highbush varieties should have any drooping branches removed, as well as removing any growth that crowds the center of the bush. Also remove any damaged, diseased, weak, or dead material.

Lowbush blueberries should have all branches cut to ground level every two to three years. These won't produce fruit the following year after pruning, so if you have lots of plants, stagger your pruning cycle so that you still have a harvest.

As the fruit is developing, you may want to place bird netting over your plants to prevent the local wildlife from eating it all.

Those living in colder climates should consider covering their highbush or half-high plants in the winter to protect them against harsh freezing conditions.

With a little support, raspberries are a fantastic grow bag fruit to experiment with.

RASPBERRY GROW BAG

When we think of raspberries, we usually imagine the typical red, bite-size fruits (*Rubus idaeus*), but keep in mind that there are other varieties that produce delicious yellow/gold, black, and purple berries. I'm often surprised by the sweet, typical raspberry taste of the yellow/gold varieties because they look like they should taste like something else.

There are many raspberry plant varieties that will do well in a container garden. Planting certified disease-free plants from nurseries is recommended. Keep reading for some ideas.

'Raspberry Shortcake': These bush raspberries were developed to grow in containers. The plants are compact and thornless with a rounded bush shape that's stunning in the landscape. Its easy-to-harvest red berries ripen in midsummer. 'Raspberry Shortcake' is self-pollinating and doesn't require staking, as its canes grow close together and only reach about 2 to 3 feet (0.6 to 0.9 m) in height.

'Heritage' Raspberry Bush: 'Heritage' is an everbearing bush variety that will also grow well in containers without support. 'Heritage' is the most common red variety and grows prolifically in most climates, up to 5 to 6 feet (1.5 to 1.8 m) in height. Its berries are large and freeze well.

'Red Latham': 'Red Latham' is a self-pollinating summer-bearing variety that fruits from late June to mid-July. Canes grow to 4 to 6 feet (1.2 to 1.8 m) high and produce glowing red berries. With less foliage than other varieties, it's convenient to grow in containers, but it will require staking.

Anne: 'Anne' is a self-pollinating everbearing variety that produces sweet, pale yellow berries. New canes will fruit in the fall of their first year and the early summer of their second year. Growers rave about their sweet taste and cold hardiness.

'Glencoe Purple' Thornless Floricane Raspberry: 'Glencoe Purple' is a cross between black and red raspberry plants, resulting in its lovely purple color and excellent flavor. It's a non-spreading, bushy variety that reaches less than 3 feet (0.9 m) tall, making it great for a container garden. It tolerates heat better than some varieties.

A young raspberry cane in a grow bag, putting out fresh growth for spring.

Any good, bagged potting soil will work well for these containers, although it's important to amend the potting soil with acidifying elements such as compost, aged manure, or peat moss. Compost and manure also provide essential nutrition and peat moss helps retain moisture while a balanced NPK rounds out nutritional needs. There are more details in the following fertilizing section.

Planting Raspberries

Raspberries are sold either as dormant bare-root canes or live potted plants. Bare-root canes look rather scraggly and unimpressive, and you may feel the urge to pack more than one cane into a small container. For plant health and dynamite berry production, stick to one cane per every 16-inch (40 cm) container, and several canes per every 5-gallon (19 L) container or larger.

Once you've put together the amended potting soil mixture for your container, make a hole large enough for your bare-root plant to sit comfortably without crowding its roots. The soil should cover the plant about 1 to 3 inches (2.5 to 7.5 cm) above the roots. Gently press the soil around the roots and water well. Make sure to add more soil if you notice that it settles low after watering.

The method for transplanting a live potted plant is nearly identical, except that it should be set at the same depth as it was growing in its pot.

After transplanting bare-root or live potted plants, add the stakes or trellises so you don't damage roots by adding them later. Mulch the soil surface with straw, woodchips, or similar organic material. Mulching helps control weeds and, more importantly for raspberries, preserves moisture.

Grow Bag Preparation

Growing raspberries in bags that are wide and deep will guarantee that your plants have enough space for new growth and for any stakes or trellises if support is needed. One cane would do well in a 16-inch (40 cm) pot, and if you are planting several canes, at least a 10-gallon (38 L) bag. Raspberries hate having "wet legs," so grow bags work well for them due to improved drainage.

Summer-bearing varieties need support because their canes tend to be taller and will bend with summer fruit. There are many options for supporting the canes. Depending on the shape of your container, tomato cages work well. A simpler, budget-friendly option is to press tall garden stakes into the perimeter of each container and tie twine around them at several heights for support.

Raspberry Soil Requirements

A main benefit of container gardening is the ability to control soil type and nutrients. Raspberries growing in pots require slightly acidic (pH 6.0 to 6.2), nutrient-retaining, well-draining soil. For comparison, blueberries require quite acidic soil with a pH around 4.5 to 5.5.

Sunlight and Temperature

Raspberries can tolerate partial shade, but your berry harvest will be much better if you can find full sun. That being said, raspberries are sensitive to high temperatures and do best in hardiness zones of 4 to 8. Specific varieties have been developed that thrive in zones 9 and above, so make sure when purchasing your plants that they are a good match for your hardiness zone.

Watering Raspberries

Generally, a container garden requires more water than plants grown in the ground because of exposure to and less protection from the elements. Avoid planting in unglazed terra-cotta pots, as they wick moisture away from the soil especially fast.

The key is to keep the soil consistently moist but not wet. Watering two to three times a week is usually sufficient. In windy areas, hot, dry climates or during heat waves, you may need to water your potted raspberries a couple times a day. A soaker hose can provide slow, deep watering.

Once your plants have stopped producing berries, it is no longer necessary to water regularly. If you live in an area with harsh winters, consider overwintering your pots in an unheated garage. Water the plants only enough to keep them alive during the winter months and then move them back to a sunny area after your area's frost-free date.

Fertilizing Raspberries

Adding a balanced fertilizer when preparing your soil for planting will provide a nutritional boost for your plants. When combined with compost at planting time, a powdered organic 10-10-10 fertilizer will help sustain plants for three to four months. While the plants are growing you can also supplement once or twice a month with a liquid kelp foliar spray for ongoing nutrition.

The spring after its first growing season, fertilize your container raspberries again with 10-10-10 fertilizer, once in March and again in May. Add compost to the container throughout the season and mulch the soil surface for weed and moisture control.

Pruning Raspberries

Red and yellow varieties produce new, green canes called primocanes every year. Primocanes don't produce fruit their first year. They are not mature enough to produce fruit until the second year. This is important to know for pruning and maintaining your plants.

Pruning is needed several times during the season, such as:

- In spring to clean up any damaged or diseased canes
- Midseason for size and height control
- Fall cleanup after harvest to prepare the plants for winter

Fall cleanup is the most pruning-intensive time. Prune the large green canes down to 4 to 5 feet (1.2 to 1.5 m), and cut the wimpy ones down to 1 inch (2.5 cm). Cut brown canes that have finished fruiting down to the soil line. Prune during dry weather to prevent exposure to harmful fungal diseases.

MICROGREEN GROW BAG

Microgreens are, without a doubt, the fastest-growing plants you can grow by any gardening method. For those who have small spaces, they're my favorite plant category to recommend both to new and advanced growers alike. This is due to the vast varieties of microgreens you can grow. After all, microgreens are simply normal plants sown at a much denser rate than your typical garden and harvested far earlier than you would typically harvest. They're the ultimate crop for the impatient gardener.

Make use of grow bags in the late fall through winter by growing microgreens in them indoors.

Over the years, I've grown just about every microgreen under the sun. Some are *much* easier to grow than others. This ease stems from their rapid germination and growth rate, as well as their lack of susceptibility to wilting and mold growth.

Other varieties are more difficult for a few reasons:

- **Germination:** Some seed hulls are exceptionally thick and benefit from a pre-soak before you sow them in your bag.
- **Time to Harvest:** Some crops, such as basil, simply grow slower, which means there are more opportunities for something to go wrong in the growing process.

The following microgreen planting charts are compiled both from personal experience and data from professional growers. I've split the charts into the more popular beginner microgreens as well as some rarer or specialty varieties to try if you're feeling adventurous.

On top of these practical benefits, there are real health benefits to consuming microgreens. Though studies vary about their range, it's generally accepted that most microgreen species contain nutrient levels higher than that of their mature counterparts. This may be because you're harvesting hundreds or thousands of seedlings in a single square foot, rather than allowing one plant to grow to maturity, but the verdict is clear: Microgreens are healthy . . . and they can be grown in grow bags.

COMMONLY GROWN MICROGREENS

PLANT	PRE-SOAK?	BLACKOUT TIME	GERMINATION TIME	TIME TO HARVEST
Arugula	No	4–6 days	2–3 days	8–12 days
Basil	No	5–7 days	3–4 days	8–12 days
Beet	Yes	6–8 days	3–4 days	8–12 days
Bok Choy/Pak Choi	No	3–4 days	1–2 days	8–12 days
Broccoli	No	4–5 days	2–3 days	8–12 days
Brussels Sprouts	No	3 days	1–2 days	8–10 days
Cabbage	No	3–4 days	1–2 days	8–12 days
Cauliflower	No	4–6 days	2–3 days	8–12 days
Chard	Yes	4–7 days	2–5 days	8–12 days
Chives	No	4–7 days	7–14 days	21+ days
Cilantro	No	7 days	7–14 days	21–28 days
Kale	No	3–5 days	2–3 days	8–12 days
Leek	No	4–6 days	3–4 days	12 days
Lettuce	No	3–5 days	2–3 days	10–12 days
Mustard (most varieties)	No	2–4 days	1–2 days	8–12 days
Peas	Yes	3–5 days	2–3 days	8–12 days
Red Clover	No	3–5 days	1–2 days	8–12 days
Sunflower	Yes	2–3 days	2–3 days	8–12 days
Wheatgrass	Yes	2 days	2 days	8–10 days

RARE OR SPECIALTY MICROGREENS

PLANT	PRE-SOAK?	BLACKOUT TIME	GERMINATION TIME	TIME TO HARVEST
Amaranth	No	5–6 days	2–3 days	8–12 days
Buckwheat	Yes	3–4 days	1–2 days	6–12 days
Celery Leaf	No	6–8 days	10–14 days	6 days
Chia	No	3–5 days	2–3 days	8–12 days
Corn Shoots	Yes	6 days	1–2 days	6 days
Cress	No	4–5 days	3–4 days	8–12 days
Endive	No	3 days	2–3 days	8–15 days
Fennel	No	3–5 days	3–4 days	12+ days
Kohlrabi	No	3–6 days	2–5 days	8–12 days
Mibuna	No	2–4 days	1–2 days	8–10 days
Mizuna (Red Streaks)	No	2–4 days	1–2 days	8–12 days
Mustard (Southern Giant)	No	2–4 days	1–2 days	8–10 days
Sorrel	No	3–5 days	1–2 days	10–12 days
Tat Soi	No	4–6 days	2–3 days	8–12 days

Prepare the Seeds

Select the microgreens you'd like to grow, then refer to the planting charts to see if they need any preparation. Some require a pre-soak to increase germination rate and speed. Always make sure you're working with sterilized tools and containers when growing microgreens, as mold growth is the number one issue for most growers.

Sow Seeds

When sowing seeds, a good rule of thumb is to sow at a density where no seed is more than one seed's diameter away from another seed. This is much easier to remember than a specific volume of seeds per crop, due to the different size of seeds and grow bags. Density is key when sowing, as you don't want a sparse, floppy crop of microgreens. Conversely, you don't want seeds germinating on top of other seeds, which can cause issues by restricting airflow and increasing the likelihood of mold.

After you sow your seeds, gently tamp them down into the soil, but do not cover them. Tamping down ensures they have soil contact and access to moisture without burying them.

Blackout Period

Most seeds germinate better in darkness, so because you didn't cover your seeds with additional soil, you need to provide a cover to black out the grow bag surface. I recommend adding a layer of thin burlap over the top of the bag.

If you don't have burlap, anything that adequately covers the surface will do. I've germinated microgreens with a dinner plate over the top. The added weight of the plate keeps seeds in contact with soil, and the overall strength of the germinating seeds will actually push the plate off of the surface as they grow.

During the blackout period, mist the surface of the soil once or twice a day to keep seeds moist.

Growth Period

After your microgreens have grown to about 2 to 3 inches (5 to 8 cm), it's time to remove the blackout cover and expose them to light. If they look yellow and sickly at this point, that's okay. After twelve to twenty-four hours exposed to sunlight, they'll start turning green and exploding with growth.

During this period, water deeply once per day and keep an eye on your microgreens for mold growth. Mold growth looks like spindly white filaments creeping along the surface of the soil. It's easy to confuse mold with root hairs, which are tiny fine white roots that emanate in a circle from the main root of each seedling.

Harvest

Reference the growing charts for estimated harvest dates for the crop you're growing. When you're nearing that date, take sterilized scissors or shears and clip off your microgreens about ½" to 1 inch (1 to 2.5 cm) above the surface of the soil. Sacrificing some stem is no big deal, as you also minimize the amount of dirt and seed hulls that you harvest.

You don't *have* to wash your microgreens post-harvest, but if you want to be extra sure you've got a clean harvest, I recommend a cold water dunk and spin in a salad spinner.

To store your microgreens for longest life, store in a container that has a small crack in it to let some air in. Wrap your microgreens in a barely damp paper towel. The extra humidity in the storage container will help keep your microgreens fresh for five to seven days. Of course, the best way to store them is in your stomach, as they're best enjoyed fresh!

SALSA GROW BAG

Designing a grow bag theme after a popular dish is a fantastic way to make a clear tie between the garden and the kitchen, which is the reason most of us are growing food in the first place. Salsa gardens are a classic layout in raised beds.

But, they often require at least 4 by 4 feet (122 by 122 cm) of space and a lot of time to grow everything that's "supposed" to be in a salsa garden, including:

- Tomato
- Pepper
- Onion
- Garlic
- Cilantro

To account for the fact that crops such as garlic and onions take *much* longer to grow than peppers, tomatoes, and cilantro, let's make some adjustments for our salsa grow bag:

- Determinate tomato variety
- Salsa pepper variety
- Green onions
- Garlic chives
- Cilantro

A great determinate tomato variety such as 'Glacier' will grow nice and compact, without needing robust support. Coupled with a classic jalapeño, the two biggest pieces to the salsa puzzle are already solved.

We'll replace the onions with green onions for speed of growth and regrowth potential and add garlic chives to make sure we still have a pop of that classic garlic flavor that adds so much depth to a bowl of salsa.

Finally, we'll plant cilantro *underneath* the canopy of the tomato and pepper plants for some much-needed shade. Despite its popularity in summer salsas, cilantro tends to bolt quickly in the heat. To provide extra tolerance against bolting, try 'Slow Bolt' cilantro.

Tomatoes, peppers, onions, and cilantro make up the classic ingredients of a quality homemade salsa.

A combination planting of different varieties of pothos, which will eventually trail over the edges of the bag.

HOUSEPLANT GROW BAG

Who says you *have* to grow edibles in a grow bag? Ornamental houseplants are fantastic options for grow bags and can brighten up an indoor or outdoor space. Better yet, you'll get to practice your arranging techniques in a low-risk way, as houseplants are far more tolerant of stress than your average flowerpot plants.

As a born-and-bred veggie gardener, I'll admit the world of houseplants, much less the world of ornamental arrangements, was foreign to me at the start. A friend introduced me to the simple Thriller, Spiller, Thriller mnemonic device to houseplant design that makes a pleasing grow bag every single time.

Thriller: These plants should add vertical height, draw the eye, and ideally add some color or drama to your grow bag:
- Rubber Plant
- Dracaena
- Croton
- Bird of Paradise

Filler: These plants are low growing and fill the interior of the bag with some texture and color. For fillers, consider these houseplants:
- Baby's Tears
- Scotch Moss
- Woolly Thyme
- African Violet

Spiller: Trailing plants that hang over the edge of the bag, breaking up monotony. For spillers, I go with the following options:
- Pothos
- Heart-leaf Philodendron
- String of Bananas/Pearls/Hearts
- Burro's Tail

STIR-FRY GROW BAG

If you love stir-fry as much as I do, you'll probably want to have your own "garden-to-wok" experience at some point. Asian vegetables span a wide range of fast- and slow-growing varieties, so a little advance planning is required.

The easiest way to begin with stir-fry gardening is to plan your "required" items. These are things such as garlic and ginger, the two staple seasonings that appear in virtually all Asian cuisine. Other flavoring agents you might want to grow include lemongrass, Thai basil, cilantro, and shiso leaves.

But what of the actual vegetables themselves? Root vegetables such as carrots and radishes or the tender bulbs of onions are important layers of a flavor base. Firm, snappy green beans and sweet snow pea pods add texture and crunch. Peppers, whether sweet or spicy, provide added pops of flavor. And, of course, vegetables such as broccoli, cauliflower, zucchini, bok choy, and napa cabbage are both filling and nutritious additions too.

So let's convert your unused grow bags into a garden of your stir-fry dreams.

Select Your Plants
Each grow bag should be able to contain multiple parts of your stir-fry garden, but you'll need to select plant combinations that will grow well together in a small space. A good way to begin is to consider each plant's height and growth habits.

For instance, the average bell pepper plant will reach 3 feet (0.9 m) at maturity. It will end up as the largest plant in its bag. It can't support extra weight, so it wouldn't work as a trellis for climbers such as peas or beans, but it would do well when layered with other heights of plants so that all the plants receive good sun exposure.

So you might place two bell pepper plants at one side of a grow bag. In front of them, plant some Thai basil, which is about half the height of the peppers at maturity. In front of the basil, you can place some bunching onions or green onions, as they don't provide much shade and love full sun conditions.

Plants such as broccoli or cauliflower may require a grow bag of their own due to their size. Alternatively, you can use a larger grow bag and place your largest plant in the center and surround it with other things. A broccoli's large leaves can provide shade for young carrots to grow around its base. A ring of garlic plants around that can help provide some natural protection against pests.

Cabbage, bok choy, and mustard greens grow quickly and make an amazing quick stir-fry.

Climbers such as peas or beans will need something to hang on to. Thankfully, there's an interesting solution for that. Growing baby corn, which is a favorite in stir-fries, not only will provide you with a few miniature ears of corn for your dinner, but it's also a living support structure. You won't get lots of baby corn from your stalks, but they are excellent for supporting beans. Consider pairing edamame beans with corn so that the nitrogen-fixing attributes of the beans go straight into invigorating the corn stalks.

While daikon radish is a popular addition to stir-fries, its length makes it difficult to grow in a grow bag setup. Instead, dot smaller icicle radishes among your other plants. They will happily grow in part-sun conditions and will produce nice root vegetables to add to your cooking.

A zucchini plant can also take up a sizable amount of room. By hiding baby bok choy or a small napa cabbage or two under its large leaves, though, you can provide shelter to these more tender plants. This can be incredibly beneficial during the height of summer when the sun's heat could scorch any leafy greens.

Maintaining Your Stir-Fry Vegetable Bags
With multi-cropped grow bags, you need to remain attentive. This is especially true if you've got a heavy-feeding plant or one that's prone to pests. Unfortunately, many Asian vegetable favorites fall into one or both of these categories.

If you're layering the grow bag garden so that the larger plants shade the more tender ones, pest management becomes trickier. Don't let "out of sight" become "out of mind." Be sure to lift up that shading foliage regularly and inspect the plants underneath for pest or disease issues and just to make sure they're growing well.

It's also important to consider the needs of flowering plants versus the needs of foliage plants. You can keep a floating row cover over celery or bok choy and it will continue to grow and thrive. But most fruit-producing plants need pollination, so don't forget them if they're hidden amidst your others, or you won't have your peppers or summer squash.

Try to match plants that mutually benefit by sharing the same space. A heavy feeder such as summer squash can be a great neighbor for green beans because they get extra nitrogen that the beans leave in the soil. Adding some corn into the mix will provide the support for those beans to climb above the squash, ensuring all three will perform and produce.

Apply a good-quality slow-release fertilizer for your plants to ensure they have a constant source of nutrition. If needed, you can still apply a liquid fertilizer from time to time for an added boost, but the slow decomposition of the fertilizer ensures your plants won't be starving in their tight quarters.

Seasonal Stir-Fry Selections for Storing
Depending on your local weather, you may not be able to grow all the components of your stir-fry mix at the same time. For instance, peas are often a cool-weather crop, while beans are a warm-weather crop. Many of the Brassica family of plants, such as broccoli or cauliflower, perform best in temperatures below 80°F (26.7°C), but bell peppers love the heat.

If you want to have a year-round veggie supply, you'll need to plan ahead for long-term storage. For instance, green beans or edamame can be blanched and then frozen. Their textures are still going to be good. Broccoli or cauliflower will be a little softer after freezing and thawing, but they will still taste just fine.

Practicing crop rotation through your grow bags to hit the ideal growing conditions can take a little experience. The first year is usually a bit of a learning curve. But cycling different crops through at different times of year also ensures that you'll have a consistent supply.

There will still be a few things that will have to remain seasonal unless you want to set up an indoor growing system. This often includes fresh herbs such as cilantro that are more difficult to store without losing quality. But if you store your other harvest, you can happily enjoy an abundance of stir-fries when the seasonal ingredients are ready to harvest.

AFTERWORD

I hope the ideas and concepts in this book have inspired you to go out and stretch your gardening skills. Grow bags have always been one of my favorite gardening methods, especially during my early days of rental gardening in my small space front yard, which is where I built Epic Gardening quite literally from the ground up.

They hold a special place in my green heart, infusing the text with bits of knowledge that will help you improve as a gardener, no matter what method you choose.

Keep growing,
KEVIN

RESOURCES

Companies with Offices in the United States

Smart Pots: www.smartpots.com

Grassroots Fabric Pots: www.grassrootsfabricpots.com

Gardener's Supply Company: www.gardeners.com

Empire Fabric Pots: www.empirepots.com/Empire-Fabric-Pots-_c_20.html

Bootstrap Farmer: www.bootstrapfarmer.com/pages/grow-bags-containers

Root Farm: www.rootfarm.com

RediRoot: www.rediroot.com

Gro Pro: www.gro-pro.net

Bloem Living: https://bloemliving.com/brands/bloembagz

Root Pouch: www.rootpouch.com

Spring Pot: www.springpot.com/who-we-are

Geo Pot: www.geopot.com

A.M. Leonard: www.amleo.com/leonard-grow-bags-bundles-of-10/p/VP-AGBXX

Rain Science: www.rainsciencegrowbags.com/commercial-growers

WellCo Industries: www.wellcoindustries.com/growing-supplies-1/growing-bags.html

Cherokee Manufacturing: www.cherokeemfg.com/product/easy-grow-bags

Winner Outfitters: winneroutfitters.com/collections/grow-bag

247 Garden: www.247garden.com

Vivosun: www.vivosun.com/collections/grow-bag

GrowInBag: www.growinbag.com

HonorThePlant: www.honortheplant.com/faq

RootMaker: www.rootmaker.com

Florafelt: www.florafelt.com

YieldPots: www.yieldpots.com

GroNest: www.gronest.com

EcoGardener: www.ecogardener.com

PlanetCoco (these are coir-filled bags): www.planetcococoir.com/coco-grow-bags.php

Bosmere USA: www.bosmereusa.com/search.asp?_Group=Propagation-and-Protection&_Sub1=Grow-Bags

Organic Ag Supply Company LLC: www.oasco.us/containers-grow-bags

Architec (Homegrown Gourmet line): www.architecproducts.com/#!brand/list?t-ids[]=3

Wholesale Forest Products: www.wholesaleforestproducts.com/grow-bags

IBEX Growing Systems: www.ibexgrows.com

Legacy Nursery Products: www.betterroots.com

Companies Outside the United States

Mani Dharma Biotech Pvt. Ltd.: www.manidharmabiotech.com/ldpe-hdpe-grow-bags.html

Evergreen Taurpaulin Industries: www.evergreentarpaulin.com/hdpe-grow-bags.html

AnushikA Agri Products: www.hdpegrowbag.com/maadi-thottam-organic-garden-grow-bags.html

Kohinoor Taurpaulin Industries: www.kohinoortarpaulin.in/grow-pot-bags.html

Neelgiri Taurpaulin & Co: www.neelgiritarpaulins.com/grow-bags.html

Sagar Vashist Fabric Pots Supplier: www.indiamart.com/company/109098367

ABOUT THE AUTHOR

Kevin Espiritu is a self-taught gardener who came to the craft later in life. Growing up as a typical Southern California kid with a nerdy streak in him, he began to play online poker in college as a way to pay for school. Early success in poker threw him off of the path to becoming an accountant, but after his graduation he found himself lost on what to do with his life after poker.

Falling into a deep video game addiction, the call to gardening came as a desire for a connection to the natural world. After a failed cucumber harvest, Kevin was hooked and combined his new love of gardening with his love of technology—and Epic Gardening was born.

Today, Epic Gardening is the world's largest multi-platform gardening education company, with the goal of helping 100,000,000+ people around the world learn how to grow their own food and reconnect to nature.

PHOTOGRAPHY CREDITS

Airam Morlett: Page 42

Andrea Trujillo: Pages 8 (bottom right), 40 (left)

Bob Farrier: Pages 67, 121

Bootstrap Farmer: Pages 4, 9 (middle right), 22, 146

Carrie C.: Page 9 (top right)

Cheah Chu Wen: Page 8 (top left)

David Grissom: Pages 8 (bottom left), 36

David Guhin: Page 8 (top right)

Denise Galya: Page 9 (bottom)

Emma Kennedy: Page 27 (bottom)

Gardener's Supply Company: Pages 7, 10, 24 (left), 25

Hannah Carroll: Pages 9 (top left), 27 (top left)

John Davis Hunks: Pages 30, 120, 129 (top)

Kevin Espiritu: Pages 6, 9 (middle left), 13-16, 18-21, 24 (right), 27 (top right), 28-29, 35 (right), 38, 39 (bottom left), 40 (right), 45-50, 54, 75, 76, 78 (left column), 79, 80-87, 88 (left), 90-94, 105 (left), 96-103, 104 (top), 106-111, 113-117, 126-127, 130, 132, 144 (bottom), 145, 148, 151 (top right and bottom), 152, 154, 165-171

Leticia Rita Horvath: Page 144 (top)

Liz Bacon: Page 55

M. Coronado Roman: Page 41 (left)

Mel Bartholomew: Page 32

Quarto Homes: Pages 34, 77, 124-125

Rogaine Ablar: Page 112

Shutterstock: Pages 12, 17, 35 (left), 39 (top row and bottom right), 41 (right), 43, 44, 44, 51, 53, 57-66, 70, 71, 77, 88 (right), 89, 95, 104 (bottom), 105 (right), 118-119, 122, 129 (bottom), 134, 136, 138, 147, 150, 151 (top left), 155, 156, 158, 159, 160

Stephanie Rose: Page 78 (right)

Thibaut Petillon: Page 68

INDEX